REASONING
WITH
STATISTICS

REASONING WITH STATISTICS

How to Read Quantitative Research

FOURTH EDITION

Frederick Williams

*Center for Research on
Communication Technology and Society,
University of Texas at Austin*

HARCOURT BRACE JOVANOVICH COLLEGE PUBLISHERS

Fort Worth   Philadelphia   San Diego   New York   Orlando   Austin   San Antoni
Toronto   Montreal   London   Sydney   Tokyo

Publisher	Ted Buchholz
Acquisitions Editor	Janet Wilhite
Developmental Editor	Barbara Rosenberg
Project Editor	Clifford Crouch
Production Manager	Kenneth A. Dunaway
Art & Design Supervisor	John Ritland
Cover Designer	Patty Potter

Library of Congress Cataloging-in-Publication Data

Williams, Frederick, 1933–
    Reasoning with statistics: how to read quantitative research/
Frederick, Williams.—4th ed.
        p.   cm.
    Includes bibliographical references and index.
    ISBN 0–03–053158–6 (pbk.)
        1. Statistics.   2. Social sciences—Statistical methods.
    I. Title.
    HA29.W525   1991
    519.5—dc20   91–31842
                                                                CIP

ISBN: 0-03-053158-6

Copyright © 1992, 1986, 1979, 1968 by
Holt, Rinehart and Winston, Inc.

Requests for permission to make copies of any part of
the work should be mailed to: Permissions
Department, Harcourt Brace Jovanovich, Publishers,
8th Floor, Orlando, Florida 32887.

Address editorial correspondence to: 301 Commerce
Street, Suite 3700, Fort Worth, Texas 76102

Address orders to: 6277 Sea Harbor Drive,
Orlando, Florida 32887
1-800-782-4479, or 1-800-433-0001 (in Florida)

Printed in the United States of America

2 3 4   016   9 8 7 6 5 4 3 2

REASONING WITH STATISTICS

How to Read Quantitative Research
FOURTH EDITION

Frederick Williams
*Center for Research on
Communication Technology and Society,
University of Texas at Austin*

HARCOURT BRACE JOVANOVICH COLLEGE PUBLISHERS

Fort Worth   Philadelphia   San Diego   New York   Orlando   Austin   San Antonio
Toronto   Montreal   London   Sydney   Tokyo

Publisher	Ted Buchholz
Acquisitions Editor	Janet Wilhite
Developmental Editor	Barbara Rosenberg
Project Editor	Clifford Crouch
Production Manager	Kenneth A. Dunaway
Art & Design Supervisor	John Ritland
Cover Designer	Patty Potter

Library of Congress Cataloging-in-Publication Data

Williams, Frederick, 1933–
    Reasoning with statistics: how to read quantitative research/
Frederick, Williams.—4th ed.
    p. cm.
    Includes bibliographical references and index.
    ISBN 0–03–053158–6 (pbk.)
    1. Statistics.   2. Social sciences—Statistical methods.
I. Title.
HA29.W525   1991
519.5—dc20   91–31842
                                                    CIP

ISBN: 0-03-053158-6

Copyright © 1992, 1986, 1979, 1968 by
Holt, Rinehart and Winston, Inc.

Requests for permission to make copies of any part of
the work should be mailed to: Permissions
Department, Harcourt Brace Jovanovich, Publishers,
8th Floor, Orlando, Florida 32887.

Address editorial correspondence to: 301 Commerce
Street, Suite 3700, Fort Worth, Texas 76102

Address orders to: 6277 Sea Harbor Drive,
Orlando, Florida 32887
1-800-782-4479, or 1-800-433-0001 (in Florida)

Printed in the United States of America

2 3 4   016   9 8 7 6 5 4 3 2

PREFACE

The fourth edition of *Reasoning with Statistics*, like the first three, is intended for people who wish to be knowledgeable readers of quantitative research literature in their fields. Whether or not one has the mathematical background to understand the bases of statistical methods, the practical logic underlying their application is not difficult to comprehend if approached with patience. For nearly a quarter of a century, *Reasoning with Statistics* has proven its worth to tens of thousands of students by providing an unintimidating view of quantitative research strategies. These students include many from Spain, South America, and Mexico who have studied the previous editions in translation: *Razonamiento Estadístico*, translated by José Manuel Rubio Suárez for Interamericana, S.A. de México.

At first, *Reasoning* made its mark with students who were new to quantitative methods, but over the years it has found an expanding market as a supplementary text in traditional statistics classes to help students answer the question, "What does all this mean to me?" I cannot think of a better endorsement—and a more rewarding feeling to an author—that enough students and professors have purchased and been sufficiently satisfied with this book that it is now going into a fourth edition.

I am again indebted to the many users of *Reasoning* who have made revision suggestions over the years, especially to Prof. Peter Monge of the Annenberg School of Communications at the University of Southern California, who served as a consultant for prior editions as well as for this new one.

May the fourth edition serve us all well.

F.W.
Austin, Texas
June 1991

CONTENTS IN BRIEF

Part
Six

ANALYSES OF COMPLEX DIFFERENCES AND RELATIONSHIPS

CONTENTS

Part Six   ANALYSES OF COMPLEX DIFFERENCES AND RELATIONSHIPS

14. Cluster Analysis   171

15. Factor Analysis   176

16. Discriminant Analysis   195

Tables   205

Simplified Computer Programs 210

Index 215

REASONING WITH STATISTICS

PART ONE

ON CONDUCTING QUANTITATIVE RESEARCH

1

Why Do Quantitative Research?

The reason for this book is mostly one of practicality. Although students often have backgrounds that do not include an introduction to statistical methods, they are expected to be intelligent readers of quantitative research reported in the textbooks and scholarly journals of their various areas. The aim of Reasoning with Statistics *is to serve this type of reader. This viewpoint should answer the first question typically raised by a person faced with an introduction to statistics: "Do I need some special mathematical background or talent?" The answer is no—at least not for the present approach.*

NO GUARANTEES

Just because a study has used statistics is no guarantee of its worth. In fact, statistics can be misused either intentionally or unwittingly, and it is not difficult to locate quantitative studies where statistics were not really needed at all. On the other hand, statistics can be a powerful tool for description or hypothesis testing. Probably the most valuable general skill individuals can have in statistical methods is the ability to understand the foregoing when reading research articles, books, or reports in their respective specialties. Of course, a valuable specific skill is the ability to use statistical methods in your research if you need them. *Reasoning with Statistics* can assist you on these two skill levels. On one level, and foremost, this is a how-they-do-it book, an introduction to the practical logic and application of the statistical methods used often in social science research, with examples in communication and education research. A secondary mission is to assist you in understanding how to use particular statistical methods, but you will also want to use a more traditional calculation-oriented statistics book or computer software manual as your main guide. Many references for these are given in the following chapters.

QUALITATIVE COMPARED WITH QUANTITATIVE RESEARCH METHODS

Research traditions, as well as common sense, tell us that whether or not we use statistics, we should expect a researcher to select the most appropriate overall research method for conducting a study. Practically speaking, a research method refers to the strategy, plan, and activities undertaken to accomplish the research.

Suppose, for example, that the research objective was to describe *New York Times* coverage of twentieth-century presidential campaigns. If the researcher's aim was to provide a historical account, we should expect careful scrutiny of original documents, assessment of articles relative to their temporal context, and an attempt to reconstruct the events of their time that led to the patterns of coverage. On the other hand, the research objective might be one of a more critical nature. Here we might expect a close textual analysis of the coverage, that is, an assessment of writers' intents, styles, and perhaps "themes" in their articles and editorials. Both of these approaches would be mainly qualitative in nature. They would depend upon archival evidence, original texts, and verbal accounts of the times, to name a few.

On the other hand, the research could take on a quantitative nature; that is, there might be an interest in measuring aspects of the campaign coverages. For instance, a count of column inches of articles could be used as a basis to compare relative coverage of candidates or political parties. Some type of content analysis procedure could be used to rate articles on a five-step "positive-negative" scale. Or one might look for correlations between campaign spending and press coverage. Simple statistics such as means, ranges, relative percentages, and correlation coefficients might be valuable. Even at this level, you as a reader could find yourself "reading statistics," albeit on a fairly simple level, but one in which a little knowledge on whether means or correlations are appropriately used could be very valuable.

By "appropriately," we do not mean accuracy of calculation—let us assume accuracy for now—but rather whether these results describe how the researcher has applied them. For example, if an array of column inch counts of different articles has some very high or low numbers, an arithmetic mean (total of all inches divided by the number of articles) will *not* be the best description of the typical length of the articles. Without becoming too involved in this problem now (read about measures of central tendency in Chapter 3), suffice it to say that whenever measures (including counting) are used in research—that is, the research is quantitative—it will be valuable to know a few of the concepts underlying statistical applications. Because we are beginning to use the term frequently, let us introduce a brief, working definition of *statistics:*

> *Statistics* (singular): a branch of applied mathematics that specializes in procedures for describing and reasoning from measures.

If you are studying or working in academia, you may encounter debates about the relative merits of qualitative versus quantitative methods. Frankly, as a researcher with some 30 years' experience, I find these exchanges meaningless unless they are tied to what an individual seeks to accomplish in a study. What phenomena are under scrutiny? What are the research questions or hypotheses? Is the researcher seeking simple description, prediction, or explanation? What are the research traditions of the particular line of study? And where does the researcher think he or she is going with the research and theory? Answers to these questions are the bases for selection of research methods. The choice might be qualitative, quantitative, or even both for a given research problem. Moreover, the longer that I have done quantitative research, the more I have recognized that qualitative aspects are embedded within it. My feeling is that those who totally dismiss either qualitative or quantitative research methods out of hand—and in vitro so to speak—are usually displaying their respective areas of ignorance. Let us concentrate now on a few fundamentals about quantitative methods.

WHEN QUANTITATIVE METHODS ARE APPROPRIATE

As an introduction, quantitative methods are appropriate (1) when measurement can offer a useful description of whatever you are studying, (2) when you may wish to make certain descriptive generalizations about the measures, and (3) when you wish to calculate probabilities that certain generalizations are beyond simple, chance occurrences (including application in hypothesis tests). Although these topics are introduced in detail in Chapters 3 through 5, consider the following an overview or preview.

Is Measurement Relevant and Possible?

As we have already noted, quantitative research involves gathering and reasoning from measurement, and statistics assists us in this reasoning process. In the newspaper example, the measurement was obtained from simple counting. But measurement can involve gauging and recording of other quantitative characteristics. We can rank-order, for example, as when respondents are asked to rank their 10 favorite books. This is quantitative research, although numbers are used to designate order rather than frequency (as in counting). But we make no assumption about the magnitudes between the numbers we use for ordering. That is, the difference between the first and second choices is not necessarily equal to the difference between the second and third; we only know about the order.

If we do make assumptions about intervals between numbers, we have still another type of measure. Suppose, for example, that a 10 point television violence scale was defined that had at its two extremes the two programs that most people who had been earlier surveyed considered the least and most violent programs; that is,

Least   Most

violent   violent

1 —— 2 —— 3 —— 4 —— 5 —— 6 —— 7 —— 8 —— 9 —— 10

The researcher might then ask respondents to rate additional programs between those extremes, assuming that the degrees of violence were reflected in steps (or equal intervals) of the scale. In this case, the numbers describe more than order; they are also meant to reflect magnitude.

If the topic of study can benefit from measurement, then at least some level of quantitative research may be appropriate. Sometimes uses of measurement are obvious, as in television ratings, costs of media services, educational test scores, or attitude scales. At other times, researchers may be innovative in developing measures, such as of television violence, media preference indexes, personality profiles, or speech styles.

But if measures are not apparent or if researchers cannot develop them with confidence, then quantitative methods are not appropriate to the problem. Chapter 3 gives more details.

Are There Statistical Generalizations to Be Made?

We already were introduced briefly to this in discussing the concept of the arithmetic mean, a descriptive statistic, in the newspaper example. Calculations like means allow researchers to make generalizations about an array or distribution of measures.

For another example, consider the television violence scale described above. If respondents had assigned a variety of television programs to different points on the 10-point scale, the researcher might use an arithmetic mean to describe the average violence rating for the programs relative to the extremes that were earlier noted. Within that 10-point range, descriptive statistics such as "variance" could describe how much people tended to differentiate violence among television programs. Or the researcher might show differences between groups of male and female respondents in terms of the average or dispersion of their ratings.

If the researcher has the measures upon which to calculate these kinds of descriptive generalizations and if the generalizations are useful for the aims of the research, then quantitative methods may be all the more appropriate. Chapters 3 and 4 give more details.

Are Probability and Hypothesis Tests Indicated?

Suppose a researcher had a "random" sample of 100 subscribers to cable television in Santa Monica, California. If all the cable subscribers in that small city had been given an equal chance of being in the study, then measures of that sample can be used to make certain generalizations about the "population" from which they were drawn (Chapter 4). Suppose, also, that the 100 had rated 25 of the prime time programs on that cable system according to the television violence scale described above. By calculating an arithmetic mean, the researcher might describe the average violence rating for the 100 as, for instance, 6.4. In addition, there are statistical methods that allow calculation of a given probability, say, .95, that the mean violence rating of all the Santa Monica cable subscribers is between, say, 5.8 and 7.0. These methods are used to make predictions about "populations," the topic of Chapter 5.

As you will see later, perhaps one of the most valuable applications of statistics is in making hypothesis tests. Although there is much more to this than is necessary to go into in this introductory chapter, let us consider a simple example. Suppose that the same researcher had obtained similar ratings of television violence from the cable subscribers of nearby Westwood, California, and that their average rating for violence was 7.2. Is this a big enough difference to conclude that Westwood and Santa Monica cable subscribers ascribe different average levels of violence to their prime time cable programming fare? (We are assuming that the same programs are available to both.)

With some of the same probability methods as used above to estimate the population mean, the researcher can perform a hypothesis test to answer the above question. This requires setting up a "null" hypothesis of no difference, that is, that the average ratings of television violence of the two communities of subscribers are not different from one another. An appropriate statistical test allows the calculation of the probability that two means do or do not come from the same population, which, in turn, allows for the acceptance or rejection of the null hypothesis. A typical level of probability for rejecting this hypothesis is set at about .05. For example, a calculation might show that the probability of the two means coming from the same population is .04. Because this is less than .05, the null hypothesis may be rejected in favor of its logical alternative, namely, that the populations have means that are different from one another. Could the researcher be making a mistake here? The statistical calculation estimates that the chances are less than four in 100 that if these populations were the same, the means might differ due to chance alone. Thus, there are good grounds for interpreting a difference.

If this decision procedure is difficult for you to grasp at this moment, do not worry. The important point for now is that statistical methods allow researchers to test hypotheses about group differences or similarities. Chapters 6 and 7 discuss these statistical methods and underlying theories in detail, and you will probably have no difficulty with the material after reading those chapters. (If you do, however, perhaps, then you can get your instructor or a consultant to lend a hand to the explanation.)

THE VALUE OF A CRITICAL EYE

No matter how intricate a quantitative research design might be, it is still usually important that the study "makes sense" to you. Sometimes researchers become so carried away with their methods that they lose sight of their object of study. One common failing is that researchers leap to the use of a particular measure or scale without carefully considering its underlying assumptions. The television violence scale mentioned earlier makes an important assumption that the scale's distances and their equality between points are "psychologically valid." How do we know that respondents actually have distances or magnitudes in mind when they use such a scale? Although there are methods for determining this (Chapter

3), sometimes critical assumptions of this nature are overlooked by eager researchers. Especially in the early stages of researching a phenomenon—for example, peoples' attitudes about personal computers—we know too little to be able to use traditional attitude scales. It may be important to conduct interviews, to hold focus groups, or to use other more qualitative or interpretative methods of inquiry as a prelude to developing measures.

The representativeness of individuals being studied is a long-time shortcoming of quantitative studies, not because of statistical methods but because some conveniently accessible groups were used. For example, generations of college sophomores have served in literally hundreds of studies of the psychology of attitude change. Many of these studies are quite clever and validly theoretical in pursuing differences in message forms and the like. But the problem with such studies is that they reflect a population of college sophomores, not necessarily the most interesting (and certainly not the most representative) segment of society! Context is sometimes a problem, too, as when an experiment on television viewing is done in a university laboratory rather than in a typical home-viewing situation. Will generalizations about behavior in a laboratory (perhaps where a student is getting extra course credit) carry over to a comfortable evening on a living room couch? Probably not; the convenience of doing research where one has the tools may forfeit the generalizability of the study's findings.

Granted that the foregoing problems are related more to overall research design than to statistical methods, they are often found in conjunction with quantitative studies. Another concern to bear in mind is whether a study has had the benefit of review by experts. If you are new to the world of research, you should know that most academic journals develop reputations for particular lines of inquiry. In most, articles are evaluated by peers who recommend rejection, acceptance, and often make suggestions for revisions. You can usually assume that quantitative articles appearing in such journals will reflect appropriate uses of statistics, although in some cases, there will always be honest differences of opinion.

If you are not a specialist in quantitative research, your main challenge in reading articles from well-regarded journals will be to understand the "logic" by which statistics leads the author to certain conclusions. If, for example, an analysis of variance shows that three versions of a television commercial for toothpaste lead to different buying decisions, what kind of statistical logic underlies that conclusion? What kind of reasoning allows that statement to be ascribed to the populations involved? Again, we are not talking about accurate mathematical calculations but the appropriateness of reasoning from the particular statistical model as applied in the study. It is this practical logic that has been the focus throughout the four editions of this book.

Where you need to be leery of statistical uses is in little-known journals, book chapters, unpublished works such as theses and dissertations, or conference papers in which the research, often in a preliminary form, may not have been subjected to critical reviews. In such cases, it is best to examine carefully the methods involved; it does not take a Ph.D. in statistics to spot a poorly designed study. As I have told countless classes of new students over the years:

Just because a study has numbers doesn't mean it's right;
always read with a critical eye.

SUMMARY

Quantitative research is just one among many methods of research. The main distinguishing characteristic is that measurement is used, and various mathematical formulas and procedures, called statistics, are used to assist in making generalizations or in testing hypotheses. Quantitative research methods are seen as no better (or worse) than qualitative counterparts; mainly, the criterion for determining which to use is appropriateness to the subject or phenomena under study. Some researchers may use both methods in a simple study. Quantitative methods are typically appropriate when (1) measurement is relevant and possible, (2) statistical generalizations may be applicable to the problem, and (3) when probabilities or hypothesis tests are thought useful. Finally, the use of numbers in research guarantees nothing; it remains important for the reader to make sense of and evaluate the study. Read with a critical eye!

SUPPLEMENTARY READINGS

Cook, Thomas D., and Donald T. Campbell, *Quasi-Experimentation*. Chicago: Rand McNally, 1979. A classic for communication and education researchers.

Howell, David C., *Fundamental Statistics for the Behavioral Sciences*. Boston: PWS-Kent, 1989. See Chapters 1 and 2.

Kerlinger, Fred N., *Behavioral Research*. 3rd ed. New York: Holt, Rinehart and Winston, 1986. Chapter 1 is an excellent introduction to research in the social sciences.

Kidder, Louise H., *Selltiz, Wrightsman and Cook's Research Methods in Social Relations*. 4th ed. New York: Holt, Rinehart and Winston, 1981. This is an updated edition of a classic introduction to social sciences research. Excellent for students interested in survey methods.

Williams, F., R. E. Rice, and E. M. Rogers. *Research Methods and New Media*. New York: Free Press, 1988. Gives overview of empirical research assumptions and alternative approaches.

2

Statistics and Research

In this chapter, we focus on some of the specific roles of statistics in research plans. Mainly, we are concerned with research in which observations are made. That is, the researcher actively observes and records information from behavior. More formally, this is called empirical research. If a formal plan is involved—and one which other researchers can replicate—the research is considered to be scientific, at least within the practices of social science. In this chapter, we review the typical plan for this type of research and the roles of statistics within it.

THE ROLES OF STATISTICS

As "Scientific" Research

Contrary to a commonly held attitude among some beginning students, research does not have to involve use of statistics in order to be *scientific*. Mainly, the scientific method places stress on gaining knowledge through the processes of systematic observation and of reasoning from those observations. In the social sciences, we seek a knowledge of the generalizations that can be made about society or behavior; we seek a systemization of this knowledge. And we have formal processes for developing it. Presumably, any competent social scientist could replicate the research and come to the same conclusions. We can see the reflection of these processes in the typical outline of the report of scientific research.

1. *Problem:* a precise statement of what knowledge was sought and why it was sought.
2. *Method:* the plan of the research, that is, how the knowledge was gained.
3. *Results:* a precise statement of the knowledge that was gained.

If these steps seem formal, they are meant to be, for they stress the systematization of the scientific approach. Practically speaking, we can think of statistics as providing tools that we can use in carrying out the kind of observational research that is essentially *quantitative*—that is, as described in Chapter 1, we are using applied mathematics to assist in the research process.

Observation and Statistics

One way to see how statistics enters the research picture is to consider the relationship between what the researcher observes and how the researcher reports these observations by use of statistical methods. Consider first what is meant by a *phenomenon*.

Phenomenon (singular): any object or event, the characteristics of which are susceptible to observation.

In the broadest sense, we use the word *phenomenon* to designate whatever segment of reality is under observation. For example, we might be concerned with the effects of a certain type of newspaper article on certain types of readers. We could designate this overall situation as a complex situation of phenomena (plural), or we might choose to divide it into each constituent phenomenon, for example, characteristics of the article on the one hand and characteristics of the readers' reactions on the other.

Given the designation of what is under study, the concern is usually with how characteristics of phenomena vary and how these phenomena affect or relate to other phenomena. When we speak of the varying characteristics of phenomena, we use the term *variable*.

Variable: an observable characteristic of an object or event that can be described according to some well-defined classification or measurement scheme.

In terms of the newspaper example, we might define articles as a variable according to the topic of the news items, for instance, local, regional, national, and international. Readers' reactions, as a variable, might be defined in terms of ratings of "interestingness." Presumably, then, a study could be concerned with the relationship between the variable of news topic and the variable of interestingness.

Whereas a phenomenon or variable is what is being observed, the researcher's reports of observations are considered as *data*.

Data (plural): the reports of observations of variables.

A characteristic feature of a study involving statistics is that the data are always in the form of some type of *measurement*.

Measurement: a scheme for the assignment of numbers or symbols to specify different characteristics of a variable.

Suppose that interestingness of the newspaper articles was rated on a three-point scale, where 1 indicates high interest, 2 indicates average interest, and 3 indicates low interest. Such a scale would constitute the type of measurement used in the study. Reports of observations, the data, would be in terms of these numbers.

Thus far, we have not considered who or what would be the focus of measurement. That is, in the newspaper example, who or what would provide the ratings of interestingness? Usually, we are concerned with what is called a *population*.

Population: any class of phenomena arbitrarily defined on the basis of its unique and observable characteristics.

Notice that the above definition refers to phenomena rather than strictly to people. In statistics, the word *population* is not limited to people; it refers to the total class of whatever is observed as a part of a study. If, for example, we were concerned with the manufacture of light bulbs, we could refer to a day's production as a population. In terms of the newspaper examples, however, our population is people, who provide the ratings. But this population includes only some well-defined group of people. Other populations, for example, might be all of the subscribers to the particular newspaper being studied. Depending on the researcher's interests, then, the population might be defined as female readers over 20 years of age. As noted in the foregoing definition, whatever is defined as the research population is purely arbitrary: It depends upon the limits of the research problem.

In many cases, it is not feasible to measure an entire population; hence, the researcher settles for some portion of the population, which is called a *sample*.

Sample: a collection of phenomena so selected as to represent some well-defined population.

In the case of the light bulbs, a sample of the population might be every 10th bulb to come off the production line. For the newspaper example, the sample might be the names of 100 subscribers selected randomly from a list of each of the 6000 subscribers to the newspaper.

Once observation and measurement have been undertaken, what we have is a body of measurement data. Each datum is in numeric form and represents one observation on one item or member of a population. The numeric report itself tells us what particular characteristic we have observed in terms of our measurement scheme. The total body of data represents measurements that have been made on every member of a sample or, in some cases, every member of a population.

Reasoning and Statistics

Thus far, we have discussed statistics mainly in reference to data, that is, particular numbers that we have assigned to observations. The main role of statistics is to provide a framework in reasoning from data so that generalized statements about the data can be made. In this sense, statistics provides the mathematical models for reasoning.

Suppose that the newspaper study required a description of the average rating of interestingness of each of the four different types of newspaper articles. In this case, we would probably use a familiar statistical procedure, that is, calculation of the mean.

$$\text{Mean} = \frac{\Sigma X}{N}$$

The X in this formula is a symbol for a numeric value of a variable; it can be any numeric value. In the present case, it represents the various ratings of interestingness that have been reported. The N is a symbol for another numeric value—the number of observations made. In short, we identify X and N in the formula with variables that represent our definitions of variables in the real world. Assuming that these identifications are adequate, we then follow the operations expressed in the formula—summation (symbolized by Σ) of the values of X and then division of this summed value by N to calculate the arithmetic mean. These kinds of operations are deductive in nature; they reflect mathematical rules.

The use of statistics in reasoning from data is somewhat the same as that described for the arithmetic mean. Statistics provides a great variety of procedures, each with a pattern of mathematical deductions that provides some type of statistical conclusion. As was illustrated for the arithmetic mean, we identify variables that are symbolized in the formulas. Given a set of adequate identifications, we then reason according to the deductive operations of the formula until we reach some desired statistical result. After obtaining this result, we then reverse the process of identification by generalizing from the results of the formulas to some statement about the real world. For example, if the arithmetic mean of the interestingness of local news items is 2.5 for a sample population, we would then use this value to estimate the average interestingness of this type of item to the population as a whole.

Even a simple statistical procedure such as the arithmetic mean can serve as a reminder of what is provided when reasoning with statistics, that is, reasoning mathematically from data to make some general state ment about the data. Sometimes we apply the word *statistics* to refer to the kind of "result" that a statistical formula provides. For example, statistically derived values that reflect an average, values that represent dispersion of measurements, or values that represent other such characteristics are called *descriptive statistics*.

> *Descriptive statistics:* calculated values that represent certain overall characteristics of a body of data.

If the body of data comprises an entire population, such descriptive statistics, of course, describe that population. However, if the body of data comprises only a sample, then additional statistics are necessary to make inferences about a population. These statistics are called *sampling statistics*.

> *Sampling statistics:* calculated values that represent how sample characteristics are likely to vary from population characteristics.

Suppose again that the mean of the ratings of interestingness was 2.5. If we had measured the entire population, this would be the mean for the population. But if this were the mean only of a sample, we would have to use a sampling statistic as an estimate of the likely position of the population mean. Similar to the example given in Chapter 1, such a statistic, when calculated, might tell us that if we considered the population mean as falling between 2.3 and 2.7, we would be risking about a 5 percent chance of error.

In sum, when we consider statistics in terms of reasoning from observations, there are two types of statistics, each with different roles. First, there are statistics that provide descriptive statements about bodies of data. Second, there are statistics that provide bases for estimating population characteristics, based on our knowledge of sample characteristics. In many kinds of research, these two types of statistics go hand in hand.

EMPIRICAL RESEARCH PLANS

Another way of viewing the practical roles of statistics is to view them within the context of the typical quantitative research plan. Earlier, we noted that research procedures were typically reported in terms of the *problem, method,* and *results.* Let us look at each of these aspects of the research plan in more detail to appreciate how statistics may relate to each.

The Problem

Research typically begins with some need for information. Generally speaking, this need is expressed as a rationale leading up to the particular problem under study. The problem itself is then defined as a statement of the problem. It might be expressed as either a *purpose* or as a *question;* for example:

> The purpose of this study is to determine the average amount of television viewing of all freshman students at. . . University, (or)

> What is the average amount of television viewing of all freshman students at . . . University?

Another type of problem statement is a *hypothesis*, a statement susceptible to testing by reasoning from observations. For example:

> Freshman students engage in more television viewing than do sophomore students at. . . University.

A problem is stated in hypothesis form only when there are sufficient reasons to make a prediction, for instance, a prediction based on theory, a prediction based on prior observations, or both.

The statement of a given problem always has an effect on the type of statistics to be used in carrying out the study. Consider, for example, the preceding problem statements. In all, some particular type of measurement is implied; that is, viewing time will be defined in terms of hours per day or programs per day. Also, explicit in the purpose and question statements is the concept of *average*. This means that some statistically derived group value will necessarily be computed. This need for such a group value is implied in the hypothesis statement; it might be either an average or a total.

Notice also that the problem statement also provides a preliminary definition of the population to be studied. This also will affect the type of statistics to be used. Perhaps the researcher will measure entire populations—and will need only descriptive statistics—or the researcher may sample from the population and, thus, need both descriptive and sampling statistics.

The Method

Usually, it is convenient to consider the method in terms of (1) the particular type of research method; (2) the subjects (persons being studied), materials, and procedures; and (3) the statistical analyses.

1. Method Types. Solving the research problem first involves an overall plan for gathering data. From a broad viewpoint, plans generally fall into two main categories. The distinguishing feature is whether observations are conducted without attempting to manipulate the variables under study or whether the researcher imposes particular manipulations upon the variables being studied and then observes the consequences. These two approaches are defined as follows.

Descriptive method: a research plan undertaken to define the characteristics or relationships, or both, among variables based on systematic observation of these variables.

Experimental method: a research plan undertaken to test relationships among variables based on systematic observation of variables that are manipulated by the researcher.

The descriptive method, or research involving observation but no direct manipulation of variables, is also classified under a variety of other labels, for example, *empirical, survey, normative, analytic,* or even *clinical.* Some typical types of descriptive research in the area of communication include broadcast ratings, readership surveys, studies of message content, and public opinion polling. Descriptive research in education could involve assessment of curriculum needs, the use of new course materials, teachers' uses of computers, and the like. The central feature in all such research is that some existing situation is being studied.

The experimental method, by contrast, is involved when the researcher is actually manipulating the variable or variables under study. The usual intent is to test a hypothesis of cause and effect. That is, do manipulations on one variable lead to consequences in another variable? For example, a hypothesis could be that teaching reading with Method A will lead to higher reading scores than Method B. Customarily, the two variables are distinguished as the *independent* (for example, teaching method) and *dependent variables* (for example, reading scores).

Independent variable: a phenomenon that is manipulated by the researcher and that is predicted to have an effect on another phenomenon.

Dependent variable: a phenomenon that is affected by the researcher's manipulation of another phenomenon.

Suppose, for example, that we were to test the following hypothesis:

Students taking English I by instructional television make better grades in the course than do students taking a nontelevised version of the course.

If we were to use the experimental method to test this hypothesis, part of the research strategy would entail setting up the two versions of the English I course as an independent variable. The effect to be studied, or the dependent variable, would constitute the final grades in the courses, measured the same way for both sections. In other words, the study would center on the presumed causal relationship between the way a course is taught (independent variable) and its consequences of the final grades in the course (dependent variable).

Whether a study is descriptive or experimental in design has a number of consequences on the statistical models to be used. In a descriptive study, we would, of course, expect to find descriptive statistics being used, and if sampling is involved, these would be augmented by the use of sampling statistics. The experimental method, on the other hand, involves a slightly different consideration in terms of how the two types of statistics are used.

In carrying out the experiment described above, we would select students for the televised course and the nontelevised course from *the same population*. That is, prior to the administration of the courses as an independent variable, we would assume that both classes had the same potential for achieving a particular distribution of grades in the course. This is a crucial assumption because if the students' grades in the two courses differ after completing the respective classes, we would want to reason that the difference was due to the way the courses were taught rather than to a group difference that was in existence before the study began.

After completion of the courses, the dependent variable is assessed, and the question is asked: Do these two groups of students now represent the same population in terms of grades in English I? Descriptive statistics will provide the bases for characterizing the two groups in terms of average grades and dispersion of grades. But in this case, we shall have a special use for sampling statistics. They will provide a basis for deducing whether it is likely that the two groups, as measured after having taken the course, still represent the same population in terms of English I grades. If it is found likely that the two groups do not now represent the same population, we have a basis for concluding that the courses affected them differently. Much more will be said on this particular use of statistics in a subsequent chapter on hypothesis testing (Chapter 6).

2. *Subjects, Materials, and Procedures.*   We now are concerned with who is measured, what tools are used for measurement and for conducting the study, and precisely how the tools are applied to the population being studied.

In most types of communication studies, measurement centers on each individual as a member of a population being studied. Typically, we call this person a *subject,* usually abbreviated as S. From a statistical standpoint, we are interested in Ss because, as minimal units of observation, the number of Ss defines the number of observations made in a particular study. We shall need to know this number in order to calculate most descriptive statistics; it further enters either directly or indirectly into our calculation of the various sampling statistics. We identify the number of observations included in a study as the variable N, which is included in statistical formulas.

The *materials* of a study include whatever tools the researcher has used to carry out the investigation. In an experiment, materials include whatever was used to manipulate the independent variable, for instance, the materials used in the televised and nontelevised versions of English I. Also, we often think of materials in terms of the particular measurement instrument used for the study. For example, in studying television viewing time, the materials might include some type of questionnaire. Or, in the experiment with English I, the materials would include the final examination, which provides the measure of the dependent variable. From a statistical standpoint, the most usual concern with the materials of a study is the type of measurement that has been involved. Because this is the topic of Chapter 3, suffice it to say that materials may offer a variety of different types of measurement and that the choice of particular statistics depends in part on the type of measurement used.

The *procedures* of a study refer to the precise manner in which the materials have been applied to the Ss and how this has led to the data of the study. Particular procedures carry implications concerning the use of statistics in a study. One example would be obtaining repeated measurements on the same Ss; that is, each S serves as a basis for more than one observation in a study. In such a case, this information must be used in selecting and applying some of the different types of statistical procedures that might be brought to bear.

3. *Statistical Analyses.*　Part of the task of designing a study involves deciding beforehand just what statistical procedures will be used and what criteria will be involved in reasoning from the statistical results back to the population under study. In a more conceptual sense, these decisions reflect the kinds of identification to be made in moving from variables being observed in the real world to variables symbolized in the statistical models and how, given the statistical results, identification will be made from the model itself back to the variables in the population. The selection of statistical models depends, of course, on what kinds of mathematical deductions the research plan calls for. These involve consideration, among other things, of what population characteristics are of interest (for example, averages and dispersions), what population comparisons might be made, and what type of measurement scale is involved.

Although they will be discussed in greater detail in Chapters 5 and 6, there are often criteria that must be established in deciding how to interpret the statistical results of a study. Recall, for example, what was said earlier about risking only 5 percent chance of error when estimating the location of a population mean. Ideally, the tolerance for such error is a decision that should be made as a part of the original design of a study. In other words, in setting up the study, the researcher would plan to estimate the population mean within limits associated with only 5 percent chance of error—or the researcher might not want to risk as much error and, thus, set some other level, for instance, 1 percent, depending on the nature of the study.

The foregoing decision also plays a central role in interpreting the results of an experiment. Recall the question posed in interpreting the study described above: Do the two groups, after undertaking different versions of the English I course, represent the same population in terms of grades in the course? The statistical results in this case will have to have an associated level of error, the determination of which is part of the researcher's task in designing the experiment. We shall see in Chapter 6 how this probability of error is incorporated into the logic of hypothesis testing.

A final consideration of the statistical analyses are the calculations themselves. In a practical sense, these refer simply to undertaking whatever calculations are defined by the statistical formula. In a more conceptual sense, these refer to how one follows the deductive mathematical rules and operations of the statistical model.

The Results

Generally speaking, the results of a study begin with whatever has been deduced by use of the statistical procedures. The task is then to interpret these statistical results in terms of what we identify them with in the population under investigation.

In terms of the role of statistics, the consequences of two types of logical patterns are reflected in the results of a study. On the one hand, the statistical results are themselves a product of a mathematical deduction. But when these results are identified with the real world, what we have in terms of the overall research strategy is a result that has been obtained inductively. That is, we have reached some generalization by means of observing a variety of particular instances of the variable under study—an inductive pattern. In accomplishing this overall inductive pattern, we have used mathematical deductions to aid us in the process.

This foregoing distinction in patterns of reasoning is important when evaluating the results of a given piece of research. As will be seen in subsequent chapters, statistics provides a rigorous basis for reasoning from data to what we have called *statistical results*. If phenomena in the real world have been adequately identified within the statistical model and if the model has been used properly, we can have confidence in statistical conclusions. However, the researcher still has the burden of reasoning from the statistical results to the real world. The criteria for evaluating this type of reasoning refer to the overall plan for research—its inductive adequacy—rather than to particular statistical models. In more practical terms, there are different criteria for evaluating the statistical aspect of a study and for evaluating the overall plan for research. A knowledgeable reader of this type of research will understand just how reasoning with statistics is only one part—but a vital part—of an overall research plan.

SUMMARY

We can think of *statistics*, a branch of applied mathematics, as providing tools that aid in conducting scientific research. One aspect of statistics concerns how observations are reported in terms of *measurement*. The major aspect of statistics, however, incorporates procedures for reasoning mathematically from numerical data to whatever conclusions make up the goal of a study. Such procedures are often classified into categories of *descriptive* or *sampling* statistics. Statistics can also be viewed within the context of an empirical research plan, that is, in terms of the *problem*, the *method*, and the *results*, of which they form a part.

SUPPLEMENTARY READINGS

Ferguson, G. A., and Y. Takane, *Statistical Analysis in Psychology and Education*. 6th ed. New York: McGraw-Hill, 1989.

Hays, William L., *Statistics*. 3rd ed. New York: Holt, Rinehart and Winston, 1981. Excellent introduction on the use of sampling statistics.

Kerlinger, Fred N., *Behavioral Research*. 3rd ed. New York: Holt, Rinehart and Winston, 1986.

Kiess, H., *Statistical Concepts for the Behavioral Sciences*. Boston: Allyn and Bacon, 1989.

Lovie, A. D. (ed), *New Developments in Statistics for Psychology and the Social Sciences*. New York: British Psychology Society and Methuen, 1989.

Rosenthal, Robert, and Ralph L. Rosnow, *Essentials of Behavior Research: Method and Data Analysis*. New York: McGraw-Hill, 1984. Chapters 1 through 3 on the nature of scientific investigation are clear and insightful.

SPSS, Inc., *SPSS-X User's Guide*. 3rd ed. Chicago: SPSS, 1988. This is the basic manual for running SPSS statistical programs, including those for desktop computers.

Williams, F., R. E. Rice, and E. M. Rogers, *Research Methods and New Media*. New York: Free Press, 1988. Describes approaches of quantitative research.

PART TWO

DESCRIPTIVE STATISTICS

3

Levels of Measurement

We discussed in Chapter 2 how measurement serves the researcher as a numerical report of observations. In some respects, measurement can be considered a special "language" for reporting observations numerically. Measurement bridges the gap between what a researcher reports as an observation of a variable in the real world and what has been defined as a variable in a statistical model. Recall the definition of measurement given in the last chapter—a scheme for the assignment of numbers or symbols to specify different characteristics of a variable. From a more detailed standpoint, it is necessary to understand the different types of particular measurement schemes, or what we call scales.

> *Scale:* a specific scheme for assigning numbers or symbols to designate characteristics of a variable.

MEASUREMENT SCALES

Different measurement scales offer varying degrees of exactness in describing given characteristics. The symbolic language of measurement usually comprises numerals, but what these numerals signify may range from the simple identification of categories for classification of characteristics to representations of "true values" of observed characteristics. Because scales range from low to high in the amount of information signified, such scales are often referred to in terms of "levels" of measurement. Understanding differences in these levels is important for two main reasons: (1) Selection of measurement scales depends on the type of information the researcher desires to record. (2) The scale that is selected may place restrictions on what models for statistical reasoning may be applied. The four types of scales available to the researcher are usually called the *nominal, ordinal, interval,* and *ratio.*

Nominal Scale

> *Nominal scale:* the assignment of numbers or symbols to designate subclasses that represent unique characteristics.

Sometimes called the *classificatory* scale, the nominal scale is the weakest level of measurement because it denotes the least information about observations. Researchers use a nominal scale when they classify observations into mutually exclusive categories, as in dividing colors of eyes (a class) into such subclasses as blue, brown, and green. There is no intention to signify any order among the categories—as, for example, that blue eyes are of greater importance than brown eyes along some specified dimension. Different classes of observations might be identified in terms of arbitrarily assigned numbers, as in blue = 1, brown = 2, and green = 3 (or blue = 3, brown = 1, green = 2, and so on). The prime characteristic of such categories is that all observations assigned to a given category are equivalent in terms of some characteristic and that these observations differ from phenomena in other designated categories in terms of this characteristic.

The foregoing phenomena, when classified in terms of a nominal scale, are often called *categorical* variables. Numbers also come into our use of nominal scales when we count the frequency of observations assigned to the subclasses. This use of numbers, however, reflects the operation of enumeration, not the operation of nominal scaling. The operation of nominal scaling is simply the division of a given class of characteristics (for example, television programs) into mutually exclusive subclasses (for example, news, drama, cartoons, and game shows).

Studies using "content analysis" in communication or education research often use nominal scales. For example, there have been a number of studies on the frequency with which different ethnic groups are depicted in children's textbooks. Nominal classification (or scaling) has been used to designate such ethnic categories as white, black, Hispanic, Asian, and American Indian. Then, enumeration is used in counting the frequency with which the characters in textbooks are classified into the ethnic categories. Eventually, the relative frequencies of characters in ethnic categories in the textbook are compared with the relative percentages of ethnic groups in the student population.

Ordinal Scale

Ordinal scale: the assignment of numbers or symbols to identify ordered relations of some characteristic, the order having unspecified intervals.

An ordinal scale incorporates the classificatory quality of the weaker nominal scale, but the important distinction is that among the subclasses there is an interrelationship of rank ordering. That is, every subclass can be compared with every other subclass in terms of some "greater than" (or "less than") characteristic. The nature of the "greater than" depends, of course, on the characteristic under study. It may indicate such relations as "more preferred," "more emotional," or "more ethical." What ordinal scaling does not represent is the magnitude of difference between ordered categories; it only denotes order or ranking.

Questionnaires used in survey research often involve ordinal scaling. In a variety of studies, you will find such queries as:

My preschool child watches television:

(1) very much
(2) a little
(3) not very much
(4) not at all

[or]

Please rank in order below the importance to you of the following media for staying current with local news (radio, newspapers, television, magazines):

1. _____
2. _____
3. _____
4. _____

In both of the cases above, numbers are assigned to indicate the relative order of something, but there is no assumption that the difference between, say, 1.0 and 2.0 is equal to the difference between 3.0 and 4.0. Nor is there an assumption that the difference between 2.0 and 4.0 is twice the magnitude of the difference between 1.0 and 2.0. In short, nothing is said or assumed about the magnitude of intervals between numbers other than that they are unspecified. But it is assumed that 2.0 is less than 3.0, or 4.0 is greater than 3.0, and so on. The numbers are labels of order, nothing more.

Interval and Ratio Scales

Interval scale: the assignment of numbers to identify ordered relations of some characteristic, the order having arbitrarily assigned and equal intervals but an arbitrarily assigned zero point.

Ratio scale: the assignment of numbers to identify ordered relations of some characteristic, the order having arbitrarily assigned and equal intervals but an absolute zero point.

The first characteristic to note in the definitions of interval and ratio scales is that magnitude is treated in terms of known and equal intervals. A second characteristic of these scales is that they are thoroughly quantitative; they are always expressed in numbers, or representations of numbers. It is also useful to point out that interval scales meet most, and ratio scales meet all, of the mathematical assumptions necessary to perform arithmetic operations. By contrast, there are restrictions on which arithmetic operations can be applied when using nominal or ordinal scales.

There are, however, some important differences between interval and ratio scales. These differences stem from the consequences of having an arbitrary zero point (interval scale) as compared with a true zero point (ratio scale). When considering an interval scale, the focus is on the *difference* between values on the scale. Because the numbers are based on an arbitrary zero, their values may not adequately reflect the real world, nor are they isomorphic to the structure of arithmetic. However, these numbers do adequately define intervals on the interval scale; therefore, the magnitudes of difference along an interval scale provide useful designations of characteristics of the real world. Furthermore, these differences are compatible with the rules of arithmetic.

Consider, for example, the two interval scales illustrated in Figure 3.1.

FIGURE 3.1   Fahrenheit and centigrade temperature scales

These are the Fahrenheit and the centigrade scales for measuring temperature. Both are interval scales; zero is arbitrarily defined. The true zero would necessarily represent that point at which the characteristic being measured vanishes, that is, the absolute absence of heat. The locations of arbitrary zero of the Fahrenheit and centigrade scales were set for convenience, and furthermore the nature of a true zero in temperature was unknown at the time.[1]

To illustrate how we can deal with differences, but not intrinsic values, in using the interval scale to measure phenomena, consider the following: The ratio of two differences on the Fahrenheit scale should equal the ratio of equivalent differences on the centigrade scale.

Suppose that we set this ratio as the difference between two thirds and one third of the range in the temperature between freezing and the boiling point of water. If the earlier statement is true, then the following would be true:

$$\frac{212°F - 92°F}{212°F - 152°F} = \frac{100°C - 33.3°C}{100°C - 66.7°C}$$

Which, indeed, it is $(2 = 2)$. What, then, about ratios of *values?* Would ratios of values representing the two-thirds and one-third scale points between freezing and boiling be equal across the two interval scales? The answer is no, which is a consequence of arbitrary zero points.[2]

$$\frac{152°F}{92°F} \neq \frac{66.7°C}{33.3°C}$$

[1] Zero on the Fahrenheit scale approximates the temperature produced by mixing equal quantities (by weight) of snow and common salt; on the centigrade scale, zero is the freezing point of water.
[2] The symbol $\neq$ is read as "not equal to."

FIGURE 3.2   Inch and centimeter scales

Consider next the two ratio scales, inches and centimeters. Suppose that the line in Figure 3.2 is exactly 1 yard (36″ or 91.44 cm) in length. With a ratio scale, ratio of scale values and differences reflect the real world. Consider the example of the ratios of two-thirds to one-third differences; here we shall consider 1 yard as the overall range, similar to the range in temperature between freezing and boiling. What was true for differences on the interval scale is true for those on the ratio scale; for example:

$$\frac{36'' - 12''}{36'' - 24''} = \frac{91.44 \text{ cm} - 30.48 \text{ cm}}{91.44 \text{ cm} - 60.96 \text{ cm}}$$

But in terms of ratios of values, what was not true for the interval scale is true for the ratio scale:

$$\frac{24''}{12''} = \frac{60.96 \text{ cm}}{30.48 \text{ cm}}$$

To summarize, the interval scale specifies the magnitude of differences among ordered points on a scale. Such intervals, which are the relative rather than the absolute values of the numbers that define them, are applicable to the rules of arithmetic and to statements about phenomena. The ratio scale, by contrast, specifies values and differences that are applicable to arithmetic operations and to statements about phenomena.

In cases where the research problem calls for more than classification or ranking, it would, of course, be desirable to work with ratio scales. With enumeration (counting) operations, we usually work with a ratio scale. Zero is usually absolute, and numeric values are usually isomorphic with the rules of arithmetic. We say "usually" because sometimes the enumeration operation is applied to psychological tests in which, for example, the number correct is counted. Solely from the enumeration standpoint, zero correct on such a test is an absolute value. However, from a psychological standpoint, zero correct may not characterize zero ability.

In fact, the notion of zero ability may never lend itself to feasible or even to useful definition. In such cases, then, we are considering scaling rather than strict enumeration, and we are once again dealing with the properties of an interval or an ordinal scale.

In communications research, ratio scales often come into use in the measurement of physical phenomena—characteristics of objects, events, or biophysical characteristics of the human organism. A biophysical example is represented in a respirometer, which provides measures of air moved (cubic centimeters) during any specified time period. Such a measure has an absolute zero (that is, no air moved) and is comparable with alternative scales also based on this zero point (for example, cubic inches).

Unfortunately, psychological aspects of behavior do not usually lend themselves to ratio scaling. Of note, a true zero is often difficult to conceptualize and is perhaps not even useful in characterizing psychological characteristics. For example, what does zero intelligence mean? At what point does a state of encoder anxiety truly vanish? There do, however, exist psychological attributes, such as in studies of psychophysical scaling (for instance, sound perception), in which properties of a ratio scale are achieved. But when measuring most psychological aspects of behavior, the communications or education researcher is typically faced with use of an interval scale and, more often than desirable, an ordinal scale.

The fine line between ordinal and interval scaling is a recurring problem in measurements such as those used to describe psychological judgments. The frequently used "semantic differential" is a good example.[3] Suppose that individuals are asked to rate their attitudes toward the United Nations by placing a check mark somewhere between the "good" and "bad" poles of the following scale:

GOOD ____ : ____ : ____ : ____ : ____ : ____ : ____ BAD

A conservative assumption is that the scale has the properties of ordinality. That is, for example, a check mark two cells to the left of the center would indicate a greater association between good and United Nations than a check mark one cell to the left of the center. The assumption of only ordinality, however, will quite drastically restrict the researcher in terms of statistical procedures applicable to the data. A more liberal assumption is that the scale cells have equal intervals, thus achieving the properties of an interval scale. What this involves is assuming that the distances between the cells are equal and that such equality is psychologically meaningful to the user. If this assumption can be made, then a far greater variety of statistical models are available for data analysis and interpretation. No doubt, this raises a dilemma for researchers. Should they assume

[3] Charles E. Osgood, George J. Suci, and Percy H. Tannenbaum, *The Measurement of Meaning*. Urbana: University of Illinois Press, 1957.

only the power of an ordinal measurement and, thus, restrict the scope of analyses, or should they assume interval measurement and, thus, gain the use of more powerful statistical procedures for reasoning from the data? The grounds for answering this question are not simple. Indeed, the answer usually depends on a compromise between what we want to achieve in terms of having confidence in the validity of results and what we are willing to risk as error.

Scales of ordinal, interval, and ratio types are associated with *continuous* variables.

MEASUREMENT ADEQUACY

Procedures for statistical reasoning are themselves without safeguards for warning the researcher when a measurement scale leads to spurious description. Nothing in the procedures will prevent the researcher from using numbers from the ordinal scale in a statistic requiring an interval scale; the problem comes in reasoning back to something true about the real world. Accuracy in measurement is like accuracy in language—it depends on the user. Two facets of the measurement problem concern the qualities of *validity* and *reliability*.

Validity

Validity: the degree to which researchers measure what they claim to measure.

The question of validity is a question of "goodness of fit" between what the researcher has defined as a characteristic of a phenomenon and what he or she is reporting in the language of measurement. In more abstract terms, it is the question regarding the truth of the measurement. For example, to what degree do the scores of a listening comprehension test represent what we conceive to be a person's ability in listening? To what degree does the numerical coding of a check mark on a seven-cell scale represent a person's attitude? Questions such as these characterize the problem of validity of measures.

Reliability

Reliability: the external and internal consistency of measurement.

The problem of reliability raises a question somewhat different from validity. Rather than the truth of a measure, external reliability asks: If the measures were applied and reapplied under precise replication of conditions, would the same results be obtained? Closely related to this is the question raised by internal reliability: If the measure comprises subparts (as in a multiple-section test of some type), do these parts contribute to equivalent results? Reliability, then, poses an objective of consistency.

PROBLEMS OF VALIDITY AND RELIABILITY

The practical concern with validity and reliability can be illustrated in a study once conducted by the author. The research problem involved measuring the time taken by encoders who were responding to a signal to create a specified message. The signal that was used to start the encoding sequence and the signal that indicated the encoder's completion of the message were both recorded by pen markings on a paper tape calibrated in time intervals. Encoding time was defined as the duration between these start and completion signals.

What the author claimed to measure (validity) was the time taken to encode the message. From a technical standpoint, the assessment of validity required answers to such questions as: Were the start and completion signals accurately recorded as anticipated on the tape? Did the time calibration on the tape correspond to what is defined by known standards as seconds? From a more behavioral standpoint, however, the assessment required answers to the following typical questions: Did the encoder recognize the start signal as anticipated? Was the time interval between the start and completion signals always indicative of encoding behavior (and not some other behavior, for instance, picking up a pencil that had dropped)?

What the author hoped would be consistent (reliability) would be answered in response to such questions as: If the same encoding event could be precisely replicated (without practice effects), would the time measures be precisely the same?

The foregoing example also illustrates two additional points—the general sources of criteria for evaluating validity and reliability and the consequences that validity and reliability have on one another.

In order to assess validity of the time measures, it was necessary to use some type of outside standard, that is, the comparison of the measure with something other than itself. One part of this validity assessment depended on an observer verifying that the recorded signals for start and completion did, in fact, represent the events. Validation of the time measure required the use of some separate standards, in this case, a stop watch.

The main test for reliability, by contrast, was a comparison of the measure with itself. In this case, a series of constant start-to-completion durations was repeated a number of times, and the recordings were compared. Although there are many and more complex approaches for assessing validity and reliability than this illustration provides, the ideas that evaluation of validity requires some type of outside standard and that evaluation of reliability requires some way of comparing a measure with itself remain basic considerations.

A final point concerning validity and reliability is the implication that they carry for one another. Does validity imply reliability? Does reliability imply validity? The answers are yes and no, respectively. If the signals of the encoding events were properly recorded on the tape and if the time durations corresponded to an outside criterion, such validity would imply (at least at the moment) that the measures are also reliable. A fallacious assumption is that reliability implies validity. The time calibration of the equipment may be perfectly consistent with itself, but this carries no implication whatsoever that it is measuring what has been defined as a *second*.

SUMMARY

Measurement involves the application of some numeric of symbolic scheme used to designate characteristics of a variable. As such, measurement bridges the gap between what we record as observations of a variable in the real world and what we may define as a variable in a statistical model. Different measurement schemes include: the *nominal*, subclassifications based on unique characteristics; the *ordinal*, subclassifications that are rank ordered; the *interval*, ranking with known intervals but with an arbitrary zero; and the *ratio*, ranking with known intervals but a true zero. *Validity* of measurement is the extent to which the researcher measures what he or she claims to measure. *Reliability* refers to the internal and external consistency of measurement.

SUPPLEMENTARY READINGS

Howell, David C., *Fundamental Statistics for the Behavioral Sciences* Boston: PWS-Kent, 1989. See Chapter 3.

Kidder, Louise H., *Selltiz, Wrightsman and Cook's Research Methods in Social Relations*. 4th ed. New York: Holt, Rinehart and Winston, 1981. Chapter 9 contains an introduction to scaling techniques.

Siegel, Sidney, *Nonparametric Statistics for the Behavioral Sciences*. New York: McGraw-Hill, 1956. Old but not outdated. Chapter 3 presents a concise section on measurement levels and the implications for choice of statistical methods.

4

Describing Distributions

Once data have been recorded according to a measurement scheme (Chapter 3), the next step is to prepare some overall descriptions of what has been observed. In any type of inductive research, we are interested in making generalizations about what has been observed. Statistics provides us with ways by which to reason in a systematic and objective manner to various descriptions that represent generalizations about data. We then consider these descriptions as a basis for saying something about the real world.

Descriptions of data may range from simple arrangements of measurement values in tabular or graphic form to complex descriptive indexes based on certain statistical models. In one way or another, such descriptions or descriptive indexes are based on the *distribution* of measurements.

Distribution: a collection of measurements usually viewed in terms of the frequency with which observations are assigned to each category or point on a measurement scale.

TABULAR AND GRAPHIC DESCRIPTIONS

Nominal and Ordinal Measures

As indicated in the definition above, the two main ingredients in a distribution are the measurement scale and the frequency of observations assigned to different categories or points on that scale. In some cases, we are interested in distributions in which the measurement scale is simply one of classification by category, that is, a nominal scale. Figure 4.1 shows a hypothetical example of a graphic summary of data in which nominal scaling is involved. Here the categories represent messages on particular topics (for example, "cancer cures") and sources of these messages. The values listed under the heading N are tabular entries that indicate the number of people who were asked to rate the trustworthiness of the sources of the messages. The graphic portion (bar graph) indicates the percentage of these people (of N) who rated the message source as trustworthy.

Notice that the measurement categories in Figure 4.1 could be arranged in any order. Also, each bar in the graph could be replaced with any type of drawing that could represent the percentages. We might even use quantities or portions of some type of cartoon figure to represent ratings of trustworthiness (that is, a pictograph).

Many times, we are concerned with measures that represent a related series of categories along a scale. Although the categories shown in Figure

Topic	Source	N	Percent rating source as trustworthy

FIGURE 4.1   Ratings of credibility of message sources (simulated)

4.1 were not arranged in order (albeit without loss of effect), it is sometimes useful to depict the order that relates the measurement categories. Table 4.1 provides an example of a hypothetical distribution involving ordered categories. Here the measure is one of general attitude toward educational television (ETV); the scale ranges from a category labeled "very positive" to a category labeled "very negative." Again, the frequencies are reported in terms of percentages. This tabulation also shows how two distributions

Table 4.1   **Comparative Rating of General Attitude Toward ETV by Teachers with and without ETV Available (Simulated)**

Rating	With ETV N = 50 (%)	Without ETV N = 100 (%)
Very positive	25	11
Moderately positive	23	35
Mildly positive	20	31
Neutral	18	10
Mildly negative	11	7
Moderately negative	3	5
Very negative	0	1

can be presented for comparative purposes; the distinction here is schools with or without ETV.

Interval and Ratio Measures

Whereas the foregoing distributions involved categories of observations, distributions will often involve a continuous scale of particular measurement values. In many types of psychological measures, we refer to each particular measurement as a *score*. Rather than being interested in scores simply as categories or ordered categories, we are interested in them as points along a scale.

Suppose that we wanted to measure how much information was retained by persons who had been exposed to a particular message. For this purpose, consider that we had constructed a 15-question test covering the message material. In particular terms, this measure can provide any of 16 scores ranging from 0 (all wrong answers) to 15 (all correct answers), and these scores could be depicted as a 16-point scale (counting 0 as a point). For example,

```
-- | -- | -- | -- | -- | -- | -- | -- | -- | -- | -- | -- | -- | -- | -- | --
 0    1    2    3    4    5    6    7    8    9   10   11   12   13   14   15
```
(number of correct answers)

The most commonly used symbol to indicate a single score of measurement value is X. If the information-retention test were administered to 50 people, we would then have 50 identifications of X. Suppose that the following scores were obtained:

```
5  15   4   9  14  11   3   4   7   6
4  10  13  10   5   9   8   7   6   5
8   9   4   6   0   5  11   7   6   4
7   7  11   1  12   9   8   7   7   5
6  12  15   8   4   8   9   6   7   5
```

As presented in the form above, the scores make up what is usually called ungrouped data. In this form, they tell us little about the overall picture. We might find it convenient to group the data in tabular form; usually, this is done according to a *frequency table*, as in Table 4.2.

In graphic form, these same data might be presented as a *frequency polygon* or as a *histogram*, such as shown in Figure 4.2. Notice that the shape of the distribution approximates the distribution of tally marks in Table 4.2. What we have in either the frequency polygon or the histogram is a graphic presentation of the frequency that each score was observed. For both presentations, a point is plotted above each score or measurement interval that corresponds to the frequency with which the score was observed data. In the frequency polygon, the points are simply connected; on the other hand, in the histogram, a horizontal line at each point corresponds to the interval of each score.

Table 4.2   **Example of a Frequency Table**

Score	Tally	f
15	//	2
14	/	1
13	/	1
12	//	2
11	///	3
10	//	2
9	/////	5
8	/////	5
7	////////	8
6	existing//////	6
5	existing//////	6
4	existing//////	6
3	/	1
2		0
1	/	1
0	/	1

Shapes of Distributions

Simply by inspection, we can determine some of the characteristics of the overall collection of scores. For one thing, we can see (Fig. 4.2) that the most frequent scores are approximately 4 through 9 and that relatively fewer persons had extremely high or low scores. Furthermore, the gross shape of a frequency distribution is usually indicative of certain general characteristics. Consider, for example, distributions labeled A through E in Figure 4.3. Distribution A approximates the shape of the distribution we have been discussing. Compare this distribution with distribution B, which is more peaked than distribution A. This indicates that fewer individual differences were discriminated by the measuring instrument; that is, people were more alike in terms of the measurement. A distribution can also be flat, as in distribution C. A flat distribution indicates more discrimination among individual differences in terms of the measures.

If a distribution has a peak that tends to be displaced at one or the other end of the measurement scale and has a tail strung out in the opposite direction, this pattern is called *skewedness*. It is described in terms of the direction in which the tail of the curve strings out. In distribution D, for example, the tail points toward the lower scores; it is, therefore, called a negative skew. Distribution E shows just the opposite. Here the tail points toward the higher scores; it is, therefore, called a positive skew. These descriptions are sometimes confusing because a negative skew indicates that the scores are clustered (the peak) in the positive direction. For example, consider the pattern of correct answers on a test. A negative skew indicates that people tended to correctly answer the questions, and

Frequency Polygon

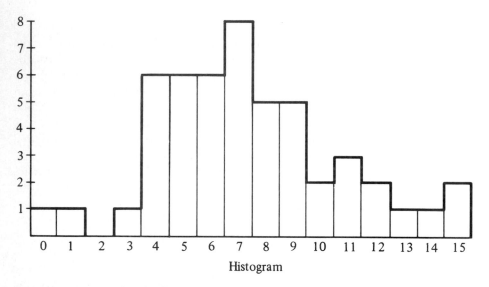

Histogram

FIGURE 4.2　Graphic displays

a positive skew indicates just the opposite—that people tended to incorrectly answer the questions on the test.

No doubt, it would be cumbersome if it were always necessary to present tabular and graphic descriptions in order to report the characteristics of a collection of measures. Fortunately, we have more economical methods

(A) Bell-Shaped

(B) Peaked

(C) Flat

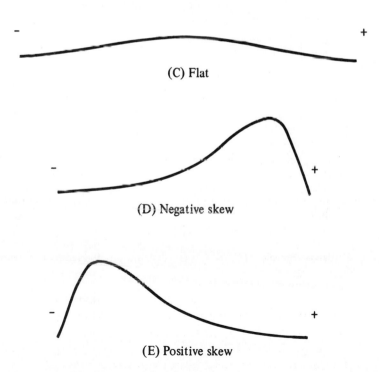

(D) Negative skew

(E) Positive skew

FIGURE 4.3   Shapes of distributions

that involve the use of statistical indexes to characterize particular aspects of measurement distributions.

INDEXES OF CENTRAL TENDENCY

One series of indexes tells us about the central tendency of the scores in a distribution. In simpler terms, these indexes refer to how scores tend to cluster in a particular distribution. When we speak of central tendency, we are really referring to an average score. Three such indexes are the *mode*, the *median*, and the *mean*.

The Mode

Mode: the most frequent score in a distribution.

Once the scores are grouped, the mode is determined by inspection. It is simply the score that has been tallied the most (see Table 4.2) or the score under the peak of the distribution (see Fig. 4.2). It is possible, however, to have a distribution in which every score occurs an equal number of times. In such a case, there is no mode. Also, sometimes a distribution has two or more scores that are of the highest, yet equal, frequencies. Such a distribution is considered bimodal, trimodal, or mul-timodal, as the case may be. All modes are then reported. For the data shown in Table 4.2 and Figure 4.2, the mode is 7. *Mo* is sometimes used as a symbol for the mode.

The Median

Median: the midpoint or midscore in a distribution.

Once scores are ordered by magnitude or are grouped, the median is defined as that point above and below which one half of the scores fall. For the data in Table 4.2, the median is 7. Of the 50 scores, 21 are greater than 7, and 21 are less than 7. If we divided the eight scores that were of the value of 7, we could define 7.0 as the midpoint of the distribution. This would then give 25 above 7.0 and 25 below, or half above and half below.[1] A symbol sometimes used for the median is *Mdn*.

The Mean

Mean: the sum of the scores in a distribution divided by the number of scores.[2]

As an index of central tendency, the mean is by far the most important. It is the most sensitive index of central tendency of the types discussed, and it is a concept incorporated in most additional statistical procedures

---

[1] If the scores within the median interval do not divide into equal groups, then the median point is determined by interpolation. Any elementary computational statistics book describes this procedure.
[2] Strictly speaking, this is called the *arithmetic mean*.

to be discussed. Most of us have calculated this type of average score long before any encounter with statistics. In Table 4.2 and Figure 4.2, the sum of the scores is 369; this is divided by 50, the number of scores, and yields 7.38, which is the mean. A commonly used symbol to denote the mean is $\bar{x}$, where the bar over the X distinguishes the mean from a value indicating a particular individual's score. There are, however, additional symbols used to signify the mean. Usually, the symbol M is used to refer to the mean of a sample, whereas the Greek lowercase letter *mu* (μ) is used to refer to the mean of a population.

It is useful for us to realize that the mean is sensitive to all of the scores in a distribution; we can think of it as a balance point. Consider, for example, the following scores:

$$8 \quad 10 \quad 13 \quad 9 \quad 7 \quad 11 \quad 10 \quad 12 \quad 10 \quad 9 \quad 11$$

If we were to locate these scores each as a single weight along a bar divided according to the measurement scale, the balance point would be the mean, as is illustrated in Figure 4.4. Given the mean, we can view the scores in the distribution in terms of their deviations about the balance point. Here (Fig. 4.4) the sum of the negative deviations should equal the sum of the positive deviations. If we take their algebraic sign into account, the sum of the negative deviations should cancel out the sum of the positive deviations. The concept of deviations about the mean as a balance point is important in certain statistical procedures and will be discussed subsequently.

Some Comparisons of the Mode, Median, and Mean

The mode and median are not necessarily balance points as is the mean. Consider the scores in Figure 4.4, for example. To change the mode (10), we would have to reduce the number of scores in the category having the most frequent occurrence or add sufficient scores in another category to create a new mode. To change the median (10), we would have to add or shift scores so as to change the half-above and half-below divisions of the distribution. In both cases, there are many scores that could be changed without any effect on either the mode or the median.

If the mean is sensitive to all scores in a distribution, why consider using the mode or median as an index of central tendency? The most practical argument favoring the mode is that it is the easiest to obtain. Once the scores are grouped, if only by frequency of occurrence, the mode can be obtained simply by inspection. If the scores are ranked, the median is also easier to obtain than is the mean. There are, however some further considerations when comparing the use of the median with that of the mean. Because of its characteristic as a balance point, the mean may sometimes be "pulled away" from what appears to be the point where the scores cluster in a distribution.

This becomes the case when there are some extremely high or low

Scores about the mean

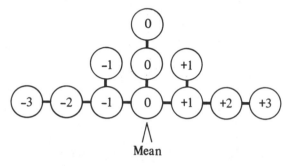

Score deviations about the mean

FIGURE 4.4   The mean as a balance point

scores in a distribution. Examine the distributions A, B, and C in Figure 4.5. Note that in distribution A, the mean is to the left of the median. This is because the few extremely low scores cause the balance point to shift to the left. Similarly, in distribution B, the mean is pulled toward the extremely high scores. The location of the mean in both distributions A and B illustrates how extreme scores in a skew tend to bias the location of the mean. In such cases, the median is usually more representative of the central tendency of the distribution—or the average score, so to speak. When a distribution is symmetrical, as in distribution C, the mean and the median are the same.

INDEXES OF DISPERSION

Just as there are indexes that describe how scores tend to cluster about the center of a distribution, there are statistical indexes that describe the dispersion or scatter across the measurement scale. The most common of these indexes are the *range*, the *variance*, and the *standard deviation*.

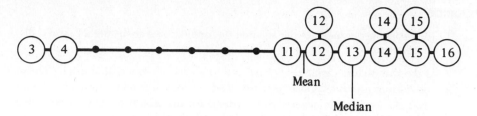

Mean

Median

(A) Mean biased toward negative skew

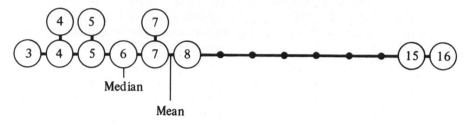

Median

Mean

(B) Mean biased toward positive skew

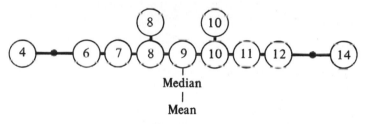

Median
|
Mean

(C) Mean and median are equal

FIGURE 4.5   Effects of skewedness upon relative locations of the mean and median

The Range

Range: the highest score in a distribution minus the lowest score.[3]

The range is the crudest index of the dispersion of a distribution. Just as the mode and the median are insensitive to all but a few scores in a distribution, so is the range. In fact, it is based on only two scores, the highest and the lowest. For the distribution in Table 4.2 (or Fig. 4.2), the range is 15. The inclusion of any number of scores between the 0 and 15 points would have no effect on this index of dispersion. There is no particular symbol commonly used to denote the range.

---

[3] Range is also sometimes defined as the *total range,* or the highest score minus the lowest score, plus 1.

The Variance

> *Variance:* the mean of the squared deviation scores about the mean of a distribution.

Like the mean, variance is sensitive to all scores in a distribution. These deviations are the same as we discussed earlier when considering the mean as a balance point among the scores in a distribution. Let us consider again the example of deviations (Fig. 4.4), but this time, we shall square each deviation value, as in Figure 4.6.

Notice first that all of the deviations in Figure 4.6 are positive; this is the result of squaring the original deviation values. We can talk about the general variability shown by these deviations simply by finding their sum (called *sum of squares;* in this case, it is equal to 30). On the other hand, we want an index of dispersion that somehow indicates an average value of the squared deviations. We obtain this by simply calculating the arithmetic mean of the sum of squares; the result is *variance* (for example, 30 divided by 11, which equals 2.73 when rounded). A lowercase s with a squared sign, s^2, and the lowercase Greek sigma with a squared sign, σ^2, are the symbols usually used to denote variance. In a strict sense, s^2 refers to the variance of a sample, whereas σ^2 refers to variance of a population.

The Standard Deviation

> *Standard deviation:* the square root of the mean of the squared deviation scores about the mean of a distribution; more simply, the square root of the variance.

Although this definition may seem confusing at first, you will probably note that it is very similar to the definition of variance. In fact, it is much easier to define standard deviation as simply the square root of the variance. This relationship is consistent with the differences in symbols used to denote variance and those used to denote standard deviation. The symbol s (that is, the square root of s^2) refers to the standard deviation of a sample, and σ refers to the standard deviation of a population. In terms of the

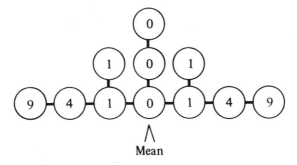

FIGURE 4.6   Squared deviations about the mean

variance (2.73) just discussed, the standard deviation would be equal to $\sqrt{2.73}$ or 1.652.

For the data shown in Table 4.2, the sum of squared deviations about the mean (7.38) is equal to 532 (rounded). This divided by the number of scores (50) in the distribution equals 10.64, or the variance. Finally, the square root of the variance, that is, the standard deviation, is equal to 3.26.

On practical advantage of using standard deviation rather than variance as an index of dispersion is that its values are smaller and easier to work with. Of more importance, however, is that standard deviation gives us a basis for estimating the probability of how frequently certain scores can be expected to occur upon the basis of sampling. We shall discuss this point in Chapter 5.

Interrelations among Score Deviations, Variance, and Standard Deviation

Because deviations from the mean and the concepts of variance and standard deviation are crucial to the understanding of many statistical models, it is important to be able to conceptualize their interrelationship. This can be done geometrically. Considering the raw data shown in example A of Figure 4.7, examine the distribution in example B of this same figure.

This distribution pictures deviations of these scores about the mean. Imagine these deviations in terms of linear distances (that is, lines that are 3 units long, 2 units long, and so on). Next, consider the results of squaring the deviations (example C in Fig. 4.7). The total area of all of the squares is equal to the *sum of squares*. When we divide this total area by the number of scores in the distribution, the average square is the result, and its area is equal to the *variance* of the distribution. The area of the average square is shown in example D of Figure 4.7. Finally, *standard deviation* can be seen as the length of one side of the average square shown in example D of Figure 4.7. (Recall that the area of a square is equal to the length of one of its sides squared.) This square provides a geometric conception of the relationship between standard deviation and variance.

DISTRIBUTIONAL INDEXES AS DESCRIPTIVE STATISTICS

In Chapter 2, we saw how each observation could be reported in the language of measurement data; in the present chapter, we saw how the collective characteristics of a given body of data could be reported in terms of selected descriptive indexes. These indexes are examples of *descriptive statistics*.

Such descriptive statistics, then, give us the capability of reasoning back to the real world in terms of what we have statistically reasoned from data. For example, we can consider statistical descriptions of a collection of data such as provided earlier in Table 4.2. If we wanted to describe the average

(A) Raw scores: 7, 13, 8, 12, 10, 9, 9, 11, 10, 10, 11

(B) Score deviations about the mean

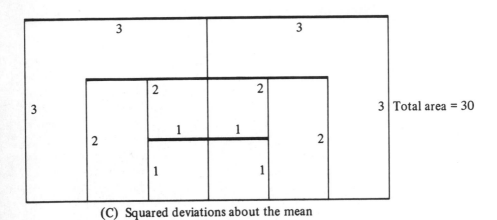

(C) Squared deviations about the mean

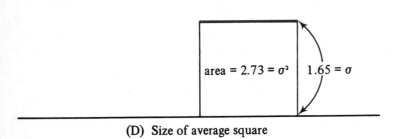

(D) Size of average square

FIGURE 4.7 Interrelations among deviations, variance, and standard deviation

level of information gain, we could turn to such descriptive indexes as
the mean, the mode, or the median. Or if we were interested in how
individuals differed in terms of this measurement, we could turn to the
range, the variance, or the standard deviation. All of these are examples
of using descriptive statistics.

USING GRAPHICS TO IMPROVE COMMUNICATIONS

The ready availability and easy use of graphics programs for personal
computers makes it feasible to use a wide variety of graphics to com-
municate effectively about data or distributions. Often, presentations to

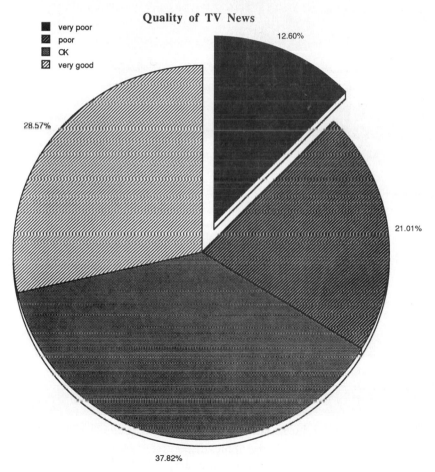

FIGURE 4.8   Examples of computer-generated graphics. Graphics are often an
effective means for conveying numerical information to a nontechnical audience.
These graphs were generated on a desktop computer.

Evaluation of Violence on Television

(Responses of 160 People)

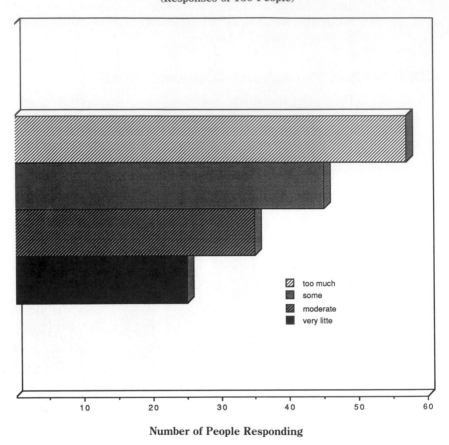

Number of People Responding

FIGURE 4.8 (cont'd.)

lay groups uncomfortable with too many numbers may be vividly enhanced by the excellent graphics displays that are now available. Although graphics can be created with dot-matrix printer characters, the best graphics are those produced by laser (or similar) printers. Figure 4.8 provides examples of graphics done with a laser printer and a desktop computer with an inexpensive graphics program.

Number of New Videogames Introduced

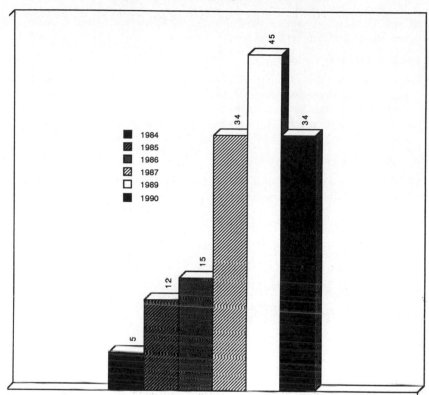

FIGURE 4.8 (cont'd.)

SUMMARY

Distributions enable us to make collective reports about a body of measurements. Such reports might be presented in terms of graphics displays. Simply by inspection, graphics displays can be interpreted in terms of how the scores cluster about a particular point, how the scores tend to be dispersed along the measurement scale, or how they may be skewed. *Descriptive statistics* provide indexes for characterizing various aspects of distributions. Indexes of how scores tend to cluster, or their central tendency, include the *mode, median,* and *mean.* Indexes of the scatter, or dispersion, of scores include the *range, variance,* and *standard deviation.* Often, the characteristics of distributions can be effectively communicated to lay readers by innovative uses of graphics.

SUPPLEMENTARY READINGS

Almost any basic statistics book provides a guide for computation of the descriptive statistics discussed in this chapter.

Two books the author has found useful are:

Hays, William L., *Statistics*. 3rd ed. New York: Holt, Rinehart and Winston, 1981. Chapter 2 discusses distributions.

Howell, David C., *Fundamental Statistics for the Behavioral Sciences*. Boston: PWS-Kent, 1989. See especially Chapters 5 and 6.

Other references include:

Lovie, A. D. (ed.), *New Developments in Statistics for Psychology and the Social Sciences*. New York: British Psychology Society and Methuen, 1989. See, especially, Chapter 1, "Simple Graphical Methods for Data Analysis."

Rosenthal, Robert, and Ralph L. Rosnow, *Essentials of Behavior Research: Method and Data Analysis*. New York: McGraw-Hill, 1984. See Chapter 16.

For computer calculations, consider:

SPSS, Inc., *SPSS-X User's Guide*. 3rd ed. Chicago: SPSS, 1988. Chapter 34 describes a computer program for calculating means, variances, and other descriptive statistics.

PART THREE

POPULATION STATISTICS

5

Predicting
Parameters

In the previous chapter, we saw how distributions and descriptive indexes of distributions enabled us to report the collective characteristics of a given body of data. In all such cases, the descriptive statistics—for example, mean and variance—were applied only to the distribution at hand; there were no attempts to generalize beyond the data to some larger population. By contrast, in the present chapter, we shall be concerned with the characteristics of populations.

INDEXES OF POPULATIONS

As discussed in Chapter 2, it is rarely feasible to observe and measure an entire population. What we do, instead, is to measure some portion (sample) that is taken as representative of a population. Then, in considering the population, two types of statistical models are brought to bear. Descriptive statistics provide a basis for calculating such indexes as the mean, variance, and standard deviation of a sample. Given these indexes, we then apply sampling statistics to estimate how these indexes apply to a population. The essence of the problem, then, is to see how we can make statements about the entire population when we have observed only a sample of that population. This can be expressed more concisely in the terminology of statistics. Consider this special usage of the word *statistic:*

Statistic (singular): a characteristic of a sample.

In the above sense, a statistic would refer to a characteristic such as an arithmetic mean, the variance, or the standard deviation of a given sample. This usage distinguishes between what we call characteristics of samples (that is, statistics) and what we call *parameters.*

Parameter: a characteristic of a population.

Suppose, for example, that we calculated the average television-viewing time of a sample of freshman students at a given university; the purpose is to make an inference about the viewing time for all freshmen at that university. The arithmetic mean of the sample would be called a statistic; this statistic is then part of the basis for estimating a parameter, that is, the estimate of the mean viewing time for all the freshman students. The relationship between statistics and parameters underlies the following definition of *statistical inference:*

Statistical inference: the process of estimating parameters from statistics.

RANDOM SAMPLING

Thus far, we have loosely considered a sample as a collection of observations that is somehow representative of a population. Let us now modify the definition of a sample given in Chapter 2.

> *Random sample:* a collection of phenomena so selected that each phenomenon in the population has an equal chance of being selected.

Quite apart from any situation involving statistics, you have probably seen random sampling involved in many different types of situations. In almost any type of survey research, various strategies will be used to try to ensure that every member in the defined population has an equal chance of being selected in the sample. Recall, too, that in the illustration of an experiment involving instructional television (Chapter 2), the two groups of subjects were drawn so that they initially would be from the same population, students in English I. This would entail random selection of the subjects from that population for the experiment, then further random assignment to one of the two groups.

In the simplest terms, we can say that if each phenomenon or unit in a population had an equal chance of inclusion in a sample, then that sample will have characteristics that can be used as a basis for estimating population characteristics. In slightly more specific terms, if random sampling is involved, this will enable us to draw upon a body of statistical theory that incorporates the mathematical relations between sample characteristics and population characteristics. Random sampling allows us to use the logic of statistical inference, that is, estimating parameters from statistics.

In order to understand better how statistics relate to parameters, it will be helpful to see the relation conceptually. This entails understanding a number of different concepts pertaining to distributions.

THREE KINDS OF DISTRIBUTIONS

One way of viewing the logic of statistical inference is to see the relation among three types of distributions—the *sample distribution*, the *population distribution*, and what is called a *sampling distribution*.

Sample Distribution

> *Sample distribution:* the frequency with which observations in a sample are assigned to each category or point on a measurement scale.

You should note that the definition above characterizes a particular type of distribution within the more general definition of a distribution given earlier. Recall the array of scores presented in Table 4.2. Suppose that now we considered them as representing a random sample of a well-defined population. As a sample, their sample distribution would be the same as the distribution presented in the foregoing chapter (Fig. 4.2), or we can

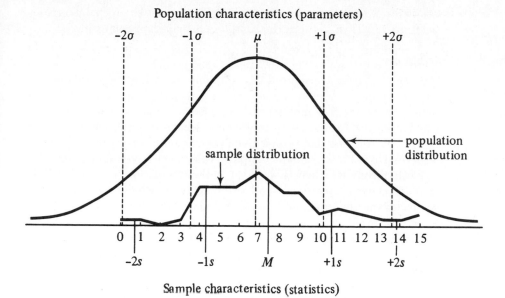

FIGURE 5.1   Comparison of sample and population distributions

view this same distribution in Figure 5.1. Notice that in Figure 5.1 we can also view certain characteristics of this sample distribution, that is, its mean (*M*) and standard deviation (*s*). In Chapter 4, we saw how these two statistics could be calculated. What we want to see next is how these statistics relate to their respective parameters, that is, a population mean and standard deviation.

Population Distribution

For purposes of illustration, suppose that we knew the precise characteristics of the population from which the foregoing sample was drawn. We could plot these characteristics as a *population distribution,* such as shown in Figure 5.1.

> *Population distribution:* the frequency with which all units or observations in a population would be assigned or expected in each category or point on a measurement scale.

Figure 5.1 illustrates a conceptual relation between sample distributions and population distributions. What we usually observe is the sample distribution. This is obtained as a result of random sampling, taking of measurements, and calculating distribution characteristics. At the same time, we assume the existence of a population distribution having as yet unknown characteristics. In statistical terms, given the knowledge of sample characteristics (statistics), we can then generalize to population characteristics (parameters). This is where the *sampling distribution* comes in.

Sampling Distribution

Thus far, we have considered distributions involving the frequency of particular units or observations as they are assigned or expected at different points along a measurement scale. Suppose now that we considered the distribution of a statistic itself. To illustrate what is meant by this, examine Figure 5.2. Imagine that from a given population we have drawn a series of successive yet independent samples. That is, we have drawn a sample, noted its mean (M), then replaced it in the population, drawn another, noted its mean, replaced it, and so on. Our knowledge of the laws of chance (or the consequences of random sampling) tells us to expect some degree of variation among the means of samples drawn from the same population. Consider examples A, B, and C in Figure 5.2; these illustrate how sample means may vary among successive samples. For purposes of reference, we have imagined for the moment that we know the population distribution. What we have, then, in these three examples is an illustration of how sample distributions can vary within a given population distribution.

Now, consider example D in Figure 5.2. Suppose that after drawing many samples, we plotted a distribution of the sample means. This is an example of a distribution of a statistic; it is also known as a *sampling distribution*.

> *Sampling distribution:* the frequencies with which particular values of a statistic would be expected when sampling randomly from a given population.

In the present illustration, we have what is known as the sampling distribution of a mean. Other statistics also have sampling distributions. We could, for example, consider the sampling distribution of s. Also, as will be seen in subsequent chapters, there are additional statistics that characterize differences between or among samples or relations among measures. We can conceive of each of these statistics as having a sampling distribution.

DISTRIBUTIONS IN TERMS OF PROBABILITY

Another way of viewing the three kinds of distributions that we have discussed thus far is to see them in terms of probabilities rather than frequencies. The problem with a frequency distribution is that it is always tied to a particular number of observations and how this number is divided among the different categories or points along a scale. Suppose that instead of considering a distribution in terms of the frequencies of units at different categories or points, we considered each frequency in terms of what proportion of the total number of units it represented. Such proportions, then, could be generalized to cases where there are different total numbers

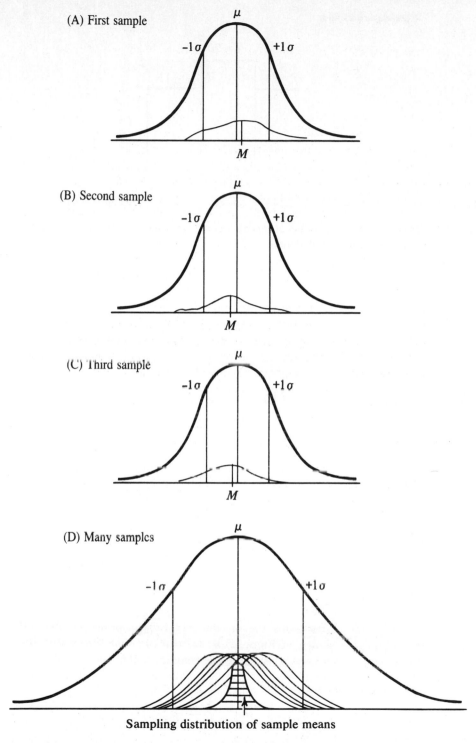

(A) First sample

(B) Second sample

(C) Third sample

(D) Many samples

Sampling distribution of sample means

FIGURE 5.2   Distribution of sample means as a statistic

Heads $\longrightarrow$	0	1	2	3	4	5	6
Probability $\longrightarrow$	.0156	.0938	.2344	.3125	.2344	.0938	.0156

FIGURE 5.3   Distribution of the number of heads expected in tossing six coins 64 times: outcomes expressed in frequencies (*f*) and probabilities

of units involved. Moreover, to make the description even more useful, we consider these proportions as probabilities. This assumes that the total probability of units is considered to be equal to 1.0 in a distribution, with the distribution of these units expressed in terms of various probabilities of occurrence at points or categories along a scale.

Comparison of a Frequency and a Probability Distribution

To use a simple example to illustrate the foregoing point, suppose that we were tossing six coins at a time and were interested in the number of heads that could be expected as the outcome of any given toss. First, we will consider what frequencies would be involved. Using a mathematical basis[1] to determine what the laws of chance would predict if we tossed the six coins 64 times, we would have the distribution shown in Figure 5.3—that is, no heads (or all tails) in one out of 64 tosses, or one head (five tails) in six out of 64 tosses, or two heads (four tails) in 15 out of 64 tosses, and so on.

As a general model of our expectations, this distribution of frequencies— that is, the particular numerical values—is based on a total of 64 tosses. To extend the description of this distribution generality to cases involving other than a total of 64 tosses and to make this description more compatible to describing expectations, we can convert this description to probabilities. Thus, as also shown in Figure 5.3, we can say that the probability of obtaining an outcome of no heads is roughly .0156, one head is .0938, two heads is .2344, and so on. The sum of the probabilities of all categories of outcomes (analogous to categories on a scale) would be 1.0.

[1] Called a *binomial expansion*.

The Normal Distribution Curve

Normal distribution curve: a definition of a particular functional relation between deviations about the mean of a distribution and the probability of these different deviations occurring.

The normal distribution curve is a theoretical distribution only. We may never find sample distributions or population distributions that precisely fit the normal curve, but in many cases, we do not go very far wrong if we assume that certain populations have this kind of distribution. Assuming that a population has this distribution, then, implies that a random sample of that population will also have a normal distribution. Additionally, the sampling distribution of certain statistics can also be assumed to have a normal distribution, for example, the sampling distribution of the mean in a normally distributed population.

A common illustration of the normal distribution curve is given in Figure 5.4. Notice its parameters—namely, a mean (μ) and a standard deviation (σ). In other words, the deviations in this distribution are expressed in σ units. Notice also how probability is incorporated into the area of this distribution. That is, the curve is divided into deviation segments, and the area of each represents some proportion of the area under the total curve. If we want to express these segments in terms of probability, we take the total area of the curve as equal to 1.0, then express each segment as either a proportion or a probability. For example, the probability of a deviation occurring between the mean (μ) and $+1\sigma$ is .3413, or between the mean and -1σ is .3413, or between -1σ and $+1\sigma$ is .6826. It is how particular deviational segments (that is, 1σ, 2σ, and 3σ) divide the area under this curve that expresses the "particular functional relation" mentioned in the above definition. In other words, the particular division

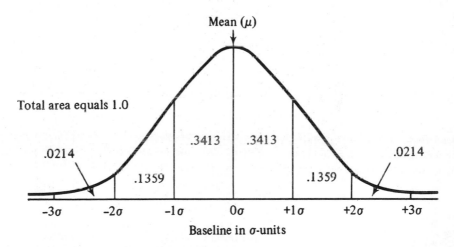

FIGURE 5.4　The normal distribution curve

of the curve into these areas according to σ segments is the characteristic quality of the normal curve.

If you summed the area segments of the curve shown in Figure 5.4, you would note that they total only .9972 rather than 1.0. What this indicates is that there are further deviations from the mean ($+4\sigma$, -4σ, -5σ, and so on) that theoretically account for the remainder of the area, or probabilities of occurrence. Being very distant from the midpoint, these deviations have extremely low probabilities of occurrence. In fact, theoretically, the tails of this curve extend out to infinity.

What we have, then, in the normal distribution curve is simply a model of a distribution having certain relations between points along the baseline (expressed in σ units) and probabilities of occurrence (areas in segments of the curve).

The Normal Curve as a Population Distribution

How do we relate the normal distribution model to actual measurement values? The tie-in comes when we identify the parameters of the normal distribution with the parameters of some actual measurement distribution. You have probably noticed that the σ symbol is the same as was used earlier for *standard deviation*. If we want to talk about such deviations in terms of the theoretical curve, we define them as σ units. In other words, the standard deviation of the theoretical curve is equal to 1.0. But if we want to talk about these same deviations in terms of scores, we find out what σ is equal to in terms of actual score units. Additionally, we assign a value of a mean to the distribution. Suppose that we know the parameters of a given population to be $\mu = 48$, $\sigma = 4.0$. Assuming that this population fits the normal curve model, then its parameters could be shown as in Figure 5.5. In terms of actual score units, the mean (μ) is now shown on the distribution as 48. That point at $+1\sigma$ above the mean is now defined as 52 (given that $1\sigma = 4.0$) or -1σ is defined as 44, or -2σ is defined as 40, and so on. What we have done is to assign the parameters (μ and σ) of a known population to the parameters of the normal curve model. We can now interpret this curve (Fig. 5.5) in terms of probabilities with which various scores would be expected. The following questions and answers illustrate the kinds of information that we can obtain from the curve.

1. What is the probability of obtaining a score between 48 and 52?
   Answer: Approximately .34. This is the same as defining the area between the mean and $+1\sigma$.
2. What is the probability of obtaining a score between 44 and 52?
   Answer: About .68, because this encompasses the area from -1σ to $+1\sigma$.
3. What is the probability of obtaining a score between 40 and 56?
   Answer: About .95, because this encompasses the area from -2σ to $+2\sigma$.

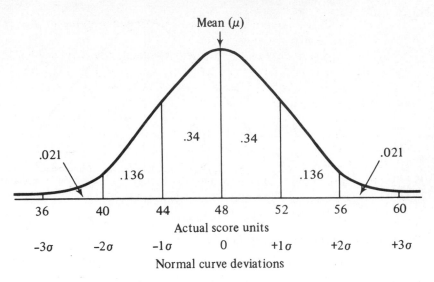

FIGURE 5.5 Actual scores in a distribution fitting the normal curve, where μ = 48 and σ = 4.0

Given the idea that we can assign different means and values of σ to the normal curve, it follows that this curve might not always be the same shape as shown in Figure 5.5. For example, as the value of each σ unit becomes larger, we would expect the curve to spread out; or as each σ unit becomes smaller, the curve would tend to pull in and have more of a peak. This point is mentioned because it is sometimes erroneously assumed that a normal distribution is always bell shaped. The central feature of the normal distribution is the defined relation between each σ unit and respective areas under the curve.

The Normal Curve as a Sampling Distribution

Recall the earlier statement that a sampling distribution is conceived as a distribution of a statistic, for example, of sample means. In example D of Figure 5.2, we saw a distribution of sample means. We can view this distribution in terms of the normal curve model. Examine Figure 5.6; this is a sampling distribution of the mean given in the previous example (Fig. 5.5). Notice that this distribution, like any other normal curve, has σ units. Only in this case, each σ_M unit refers to the deviations of sample means about the population mean. We call the σ_M unit a *standard error of the mean;* it is a sampling statistic.

> *Standard error of the mean:* the standard deviation of a distribution of sample means.

As the normal curve model is used as a basis for estimating probabilities of different scores occurring in a population, here the normal curve model is used as a basis for estimating the probabilities with which different

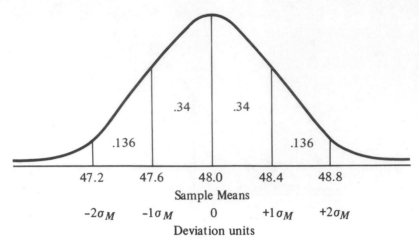

FIGURE 5.6   Sampling distribution of sample means where $M = 48$, and $\sigma_M = 0.4$

sample means will occur. The following questions and answers illustrate the information that we can gain from this curve (Fig. 5.6).

1. What is the probability of a sample mean occurring between 48.0 and 48.4?
   Answer: Approximately .34. This is the area between the mean and $+1\sigma_M$.
2. What is the probability of obtaining a sample mean between 47.6 and 48.4?
   Answer: Approximately .68, because this encompasses the area from $-1\sigma_M$ to $+1\sigma_M$.
3. What is the probability of obtaining a sample mean between 47.2 and 48.8?
   Answer: About .95, because this encompasses the area from $-2\sigma_M$ to $+2\sigma_M$.

Just as sample means have a sampling distribution, other statistics have their own sampling distributions. We have centered upon the mean because it is the descriptive statistic most commonly considered in relation to a parameter.

SAMPLING ERROR

What we have discussed thus far has illustrated the relations among sample distributions, population distributions, and sampling distributions. For example, if we have sampled randomly from a normally distributed population, we can assume that certain relations hold between what we have found as characteristics of our sample and what the characteristics of the population are likely to be. One way of illustrating this relation is in terms of the sampling distribution of a statistic—for example, the mean. A more general way of thinking about what a sampling distribution of a statistic provides is in terms of *sampling error*.

> *Sampling error:* an estimate of how statistics may be expected to deviate from parameters when sampling randomly from a given population.

In practical terms, whereas random sampling will yield sample characteristics that tend toward the population characteristics, we cannot expect sample characteristics to be precisely the same as population characteristics. The laws of chance allow sample characteristics to deviate from population characteristics, but our knowledge of these laws allows us to estimate what kinds of deviations to expect. This is the essence of statistical inference and the underlying logic of sampling statistics.

PRACTICAL PROCEDURES

In the case of actual research, we begin only with the knowledge of sample characteristics. We can then incorporate the logic discussed in the preceding sections if we can make the assumption that we have sampled randomly from the population. Additionally, if we can meet the assumptions necessary for the normal curve model, then our procedures for estimating population characteristics will be based on the population and sampling distributions fitting that model.

Let us consider again the data presented in Figure 5.1 (Table 4.2), but this time, we shall assume that we know nothing about the population, except that it is normally distributed. What kinds of statements might we make about population characteristics? Let us consider the population mean (μ) and the population standard deviation (σ).

Population Mean

We have already seen how the sampling statistic, *standard error of the mean* (σ_M), describes the likely deviations of sample means about the population mean. In making estimates of the population mean, we begin with the calculation of this statistic. One method for such calculation is shown in Table 5.1. Here the result is that $\sigma_M = .47$. Given this value, Figure 5.7 illustrates a sampling distribution of sample means estimated for this distribution, where $M = 7.38$. Thinking purely in terms of "best estimate," we could say that the mean of the sample is the best single estimate of the population mean. Usually, however, we take a more conservative approach that incorporates what we know about the sampling distribution of the sample means. Two such approaches, the first more conservative than the second, are as follows.

Table 5.1   **Examples of Calculations of Estimates of Parameters (Data from Table 4.2, Fig. 5.1)**

Sample Characteristics:

$$M = 7.38$$
$$N = 50$$
$$\Sigma x^2 = 532$$

$$s = \sqrt{\frac{\Sigma x^2}{N}} = \sqrt{\frac{532}{50}} = 3.26 \text{ (rounded)}$$

Population Characteristics:
    Standard error of the mean (σ_M)

$$\sigma_M = \sqrt{\frac{\Sigma x^2}{N(N-1)}} = \sqrt{\frac{532}{50(50-1)}} = .47 \text{ (rounded)}$$

Standard deviation (σ)

$$\sigma = \sqrt{\frac{\Sigma x^2}{N-1}} = \sqrt{\frac{532}{50-1}} = 3.295 \text{ (rounded)}$$

If we assume that the population mean lies between 6.17 and 8.59, there would be a probability of only 0.1 that we are wrong. Note in Figure 5.7 what this estimate is based on. The range between 6.17 and 8.59 encompasses .99 of the area of the sampling distribution. This corresponds to the range between $-2.58\sigma_M$ to $+2.58\sigma_M$, which leaves an area of approximately .005 in each tail of the curve. In other words, we are

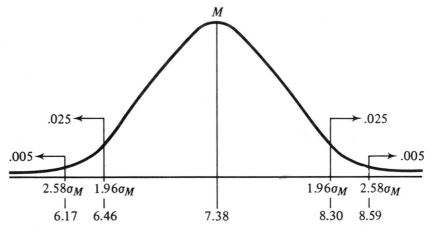

FIGURE 5.7   Sampling distribution where $M = 7.38$, $\sigma_M = .47$

incorporating estimates that are both above and below the mean of this distribution. If, by some unfortunate stroke of luck, we were indeed wrong, this would indicate that the population mean is somewhere outside of the area between $-2.58\sigma_M$ to $+2.58\sigma_M$ on the curve in Figure 5.7. You can see, however, that as the population mean might be farther and farther away from the sample mean, its chances of being so are less and less probable.

If we were willing to assume a slightly greater chance of error in our estimation, we might assume that the population mean lies between 6.46 and 8.30 and say that there is a probability of .05 that we are wrong. This estimate is within the range of $-1.96\sigma_M$ to $+1.96\sigma_M$, which leaves approximately .025 area in each tail of the curve.

The foregoing estimates reflect the types of decisions we usually make when interpreting a sampling statistic. This actually entails a hypothesis-testing decision procedure, a topic treated in greater detail in the next chapter.

Population Standard Deviation

There are a number of alternative methods for estimating the standard deviation of a population, based on a knowledge of sample characteristics. One of these is presented in Table 5.1. Recall that in Chapter 4 we noted that the standard deviation of a distribution (without any implication of its being a sample) was calculated by the formula for s shown in Table 5.1. When a distribution is considered as a sample, its standard deviation is often calculated by use of the formula shown for σ in Table 5.1; as implied by the symbol σ, this is also considered as the best estimate of the population standard deviation. If, as with the population mean, we wanted to take a more conservative approach in making our estimate, we could plot a sampling distribution of sample standard deviations. This sampling distribution would have as its standard deviation a *standard error of the standard deviation* (σ_σ), a concept analogous to the standard error of the mean. We could then take this sampling distribution and make various statements about the location of the population standard deviation and accompany these statements with estimates of likely error.

SUMMARY

The essence of statistical inference is to reason from characteristics of samples (statistics) in order to estimate characteristics of populations (parameters). We are in a position to use the logic of statistical inference when samples have been *randomly* drawn from populations. Sample characteristics are incorporated in a *sample distribution*, whereas a *population distribution* incorporates the population characteristics that are to be estimated. A *sampling distribution* characterizes the likely deviations of a given statistic about a parameter.

When distributions can be approximated to the *normal distribution curve*, we have at least one type of basis for making probability statements about parameters and likely sampling error of statistics. In a practical situation, there are various methods for calculating estimates of population characteristics, given sample characteristics.

SUPPLEMENTARY READINGS

Howell, David C., *Fundamental Statistics for the Behavioral Sciences*. Boston: PWS-Kent, 1989. See Chapter 8.
Kerlinger, Fred N., *Behavioral Research*. 3rd ed. New York: Holt, Rinehart and Winston, 1986.
Lovie, A. D. (ed.), *New Developments in Statistics for Psychology and the Social Sciences*. New York: British Psychology Society and Methuen, 1989. See Chapters 7 and 8.
Spence, Janet T., John W. Cotton, Benton J. Underwood, and Carl P. Duncan, *Elementary Statistics*. 4th ed. Englewood Cliffs, N.J.: Prentice Hall, 1983. See Chapters 2 through 4.

6

Testing Hypotheses

In the previous chapters, emphasis was placed on basic statistical concepts and mainly on how these concepts underlie our reasoning from data to descriptions of what the data are believed to represent. In the present chapter, we turn to a discussion of a type of decision procedure that incorporates reasoning with statistics. In such cases, the decision concerns a statement that we are attempting to assess by conducting a study. In short, this decision procedure concerns the role of statistics in hypothesis testing. With this procedure, we still use descriptive statistics and sampling statistics, but we use them in a rather special way.

KINDS OF HYPOTHESES

Modern science has taught us not to expect absolute certitude in statements that we make about the real world. Instead, we gauge certitude in terms of probability. In order to use probabilities for this purpose, we usually begin with statements that we wish to test in our studies; these statements are called *hypotheses*. In most psychological studies of communications or educational phenomena, however, two kinds of hypotheses are involved. One of these, called a *null hypothesis*, can be evaluated in terms of the probabilities that sampling statistics provide. The other, called a *research hypothesis*, is the actual research prediction that we want to test. The research hypothesis is always the logical opposite of the null hypothesis, so that if the null hypothesis is found to be relatively improbable (according to statistical criteria), then, by implication, the research hypothesis is considered acceptable. We need these two kinds of hypotheses in order to bridge the gap between what sampling statistics can tell us about probabilities and the kinds of statements that we want to make about phenomena.

In order to make this discussion as simple as possible, let us depart from psychological measures and use an easily visualized example that relates samples to populations. Suppose that we have scooped up two handfuls of marbles and want to know if, in terms of the average weight of the marbles, the two handfuls have been drawn from the same population, or at least populations with equal means. These randomly drawn handfuls are our examples of *samples*. The larger batch of marbles we drew them from is considered the *population*. Logically, if the handfuls were drawn randomly from the same population, we would expect the mean weight of marbles in each to be similar if not the same. Sampling error will be reflected in how much they differ if actually drawn from the same population. On the other hand, if their mean weights differ more than expected from sampling error, then we might decide that they came from different populations. In order to set ourselves up for this decision, we use the aforementioned null and research hypotheses.

The Null Hypothesis

When we make the statement that whatever differences are observed between sample means is due to sampling error, it is called a *null* hypothesis.

> *Null hypothesis:* a statement that statistical differences or relationships have occurred for no reason other than laws of chance operating in an unrestricted manner.

In this particular example, we could state the null hypothesis in terms of the phenomenon under study—that the difference between the mean weights of the two samples of marbles is due to sampling error and that they have the same population mean. Or else, we can shorten the statement in terms of a mathematical expression,

$$\mu_1 = \mu_r$$

where μ_1 is the population mean represented in the sample of the marbles in the left hand, and μ_r is the population mean relating to the marbles in the right hand.

Recall in Chapter 5 how sampling statistics can provide us with estimates of the probability that certain sample characteristics could occur by chance (based on random sampling). In the present case, this same logic applies in estimating probabilities of obtaining various differences *between* the means of the two samples. For example, we might estimate that in drawing successive pairs of samples we would expect to have a difference between the sample means as large as $\frac{1}{16}$ ounce in five out of 10 cases of random sampling (*probability* = .50), whereas the likelihood of obtaining a difference as large as $\frac{1}{4}$ ounce might only be one in 1000 cases (p = .001). In practical terms, we are able to calculate the probability that some observed difference in mean weights of the two samples could occur on the basis of random sampling. This puts us in a position to use probability as a basis for a decision about the acceptability of the null hypothesis. Sampling statistics aid us in gaining these estimates of probability.

The Research Hypothesis

Usually, however, our primary research interest is *not* in the prediction that some difference occurred by chance. Instead, we are interested in the acceptability of a statement that there is a real difference, and not one attributed to mere chance. In other words, our primary interest is in the acceptability of a statement that is the logical alternative to a null hypothesis; this statement is called the *research hypothesis*.

Research hypothesis: a statement expressing differences or relationships among phenomena, the acceptance or nonacceptance of which is based on resolving a logical alternative with a null hypothesis.

How, then, do we test the acceptability of a research hypothesis? The above definition points out the answer. The null hypothesis and the research hypothesis are logical alternatives to one another. We consider the research hypothesis as acceptable (or supported) only if we have decided that the null hypothesis is unacceptable because of its low probability. Note that the emphasis here is on the null hypothesis being considered improbable. This is a central point because our decision about the null hypothesis is the only link between what our statistical models tell us about probability and the decision that we make concerning the research hypothesis.

Suppose, again, that we are studying the two handfuls of marbles. To put the problem in a practical context, let us presume that each handful represents a random sample of the day's output of a given marble factory and that we have good reason to predict that marbles produced by the two factories (I, II) do not have the same average weight. The average we want to consider is the mean of the population—not just two handfuls. Our research hypothesis, then, might be stated as:

There is a difference in average weight of marbles in the population produced by Factory I and the population produced by Factory II.

In mathematical form, the research hypothesis would be:

$$\mu_I \neq \mu_{II}$$

The logical alternative to the above prediction would be stated in terms of the null hypothesis:

$$\mu_I = \mu_{II}$$

Sampling statistics would provide us with a basis for making a probability statement about the null hypothesis. If we found this calculation of probability to be rather low—say, .05 or less—we might use it as a basis for rejecting the null hypothesis. In other words, we would say that it seems unlikely that the difference between the two means could be due solely to sampling error. In rejecting this null statement, we would then accept its logical alternative, the research hypothesis.

A research hypothesis is either accepted or not on the basis of a decision about a null hypothesis. The decision about the null hypothesis is based on what sampling statistics tell us in terms of the probability with which the observed difference could have occurred under the terms of the null hypothesis, that is, due to sampling error. These steps will seem less cumbersome when seen in terms of the problem-method-results context of research designs.

THE PROBLEM

In terms of the research problem, a design involving hypothesis testing will incorporate a research hypothesis, a null hypothesis, and a probability level selected as a criterion for rejecting the null hypothesis.

Statement or Implication of a Research Hypothesis

The prediction of some difference or relationship may appear in many different kinds of statements, or it might not be directly stated at all in a research report. Typically, however, the statement of a research hypothesis represents a compromise between fulfilling two goals. On the one hand, the prediction is phrased in terms that will orient the reader to the problem under study; for example:

> Magazine readers and nonmagazine readers have different annual mean incomes.

On the other hand, the research hypothesis to be tested statistically must be an unambiguous mathematical statement, such as:

$$\mu_r \neq \mu_n$$

A more detailed verbal statement might be something like:

> In terms of mean annual income, there is a difference in the mean incomes of two populations as reflected by a difference between the mean (M_r) of a sample of magazine readers and the mean (M_n) of a sample of nonmagazine readers.

The crux is that a research hypothesis, however stated, must imply a null hypothesis that is susceptible to a probability estimate.

Usually, the need for precision in bridging the gap between the statement comprising a research hypothesis and what the researcher specifically means in terms of the study is served by presenting a series of definitions. In the above examples of a hypothesis, for example, precision is gained by definitions that answer such questions as: What exactly is meant by a *magazine reader* as distinguished from a *nonmagazine reader?* Are the *populations* restricted to some geographic area? What constitutes *annual income?* What is the time element involved, the period to which the statement applies?

Sometimes a researcher will use the statistical logic of hypothesis testing but will not claim beforehand to be testing specific predictions. Such cases sometimes involve a research problem stated in purpose form; for example,

The major purpose of this study was to investigate variations in the behavior of a student answering a question when negative and positive feedback are provided as a teacher's evaluation of the answer.

In the above problem statement, research hypotheses are implied in the presumed relation between the feedback and the answers. It might be stated generally as:

Verbal behavior under negative feedback $\neq$ Verbal behavior under positive feedback.

Of course, the researcher would have to define the measures involved in the prediction, the exact conditions of feedback, and so on.

There may be times, too, when the statistical implication of a research hypothesis is somewhat secondary to the researcher's general problem. It is not unusual to find problem statements in the form of questions; for example,

This research was designed to probe the general use of how people gain information about health problems. Specifically, answers were sought to the following questions:

1. Do people in the age category of 20–39 years differ from those in the 40–59 year range in the average degree to which they seek health-related information?
2. Do black families equated on a socioeconomic scale with white families differ in the average degree to which they seek health-related information?
3. And so on.

The above study does not center on testing predictions, but when conducting statistical tests such as suggested in the questions of population differences, a research hypothesis of some greater-than-chance difference is implied as well as a corresponding null hypothesis of chance difference.

Statement or Implication of a Null Hypothesis

In the previous example involving annual incomes, it was indicated that $\mu_r \neq \mu_n$, as a research hypothesis, implied $\mu_r = \mu_n$ as the null hypothesis. But suppose, for example, that the research hypothesis is:

$$\mu_r > \mu_n$$

(The reader population mean is greater than the nonreader population mean.)

The implied null hypothesis, then, is:

$$\mu_r \leq \mu_n$$

(The reader population mean is equal to or less than the nonreader population mean.)

In practice, the null hypothesis is frequently implied rather than stated. In fact, if there is nothing special about the null hypothesis—that is, it is clearly implied by the research hypothesis—a statement of the null explanation is apt to be redundant. The central feature of a null hypothesis is that it must be compatible with probability estimates provided by a statistical model and at the same time a logical alternative to the research hypothesis.

Choice of a Probability Level

Bear in mind that we are interested in selecting a probability level as a criterion for rejecting a null hypothesis. That is, if the probability calculated for the null hypothesis is at this level—or even less probable—we shall reject the null hypothesis and accept the research hypothesis.

Sometimes a probability of .05 is taken as a level suitable for rejection of the null hypothesis. An actual value of probability is then calculated by statistical procedures, and if it is at or below this .05 level, it is grounds for rejection of the null hypothesis. As a criterion for decision about the null hypothesis, this level is often called the *rejection region*, or the *significance level*.

> *Rejection region (significance level):* a level of probability set by the researcher as grounds for the rejection of the null hypothesis.

If a calculated value of probability is such that it falls within the rejection region, the researcher will interpret the difference or relationship as *statistically significant*.

> *Significance:* the level of calculated probability was sufficiently low as to serve as grounds for rejection of the null hypothesis.

Oftentimes, the level set for significance is expressed in terms of a lowercase letter p (meaning probability), a sign indicating "less than," and some particular probability value, for instance, $p < .05$, $p < .01$. Thus, if a researcher reports that a difference was "significant at the $p < .05$ level," he or she is saying simply that the null hypothesis was rejected upon attaining this level, or a lower level, of calculated probability.

There is nothing sacred about the $p < .05$ level; it is set mainly out of convention in using statistical methods. There are, however, many cases where some other rejection region will be used. Setting this rejection requires consideration of the types of error a researcher is willing to tolerate, a topic discussed subsequently.

THE METHOD

In terms of its statistical aspect, the method centers upon how we calculate values of probability associated with the null hypothesis. As mentioned earlier, this involves use of both descriptive and sampling statistics. For purposes of illustration, suppose that we were testing the hypothesis stated earlier:

Magazine readers and nonmagazine readers have different annual mean incomes.
[or]

$$\mu_r \neq \mu_n$$

We would use descriptive statistics to provide indexes of the characteristics of the samples of the two populations. Among other things, these indexes would include the two sample means, one for the readers (M_r), one for the nonreaders (M_n). We could call each of these means a *statistic* (in the sense of a sample characteristic), and we call their difference (that is, $M_r - M_n$) a statistic. What we want to know is the sampling error of the statistic, $M_r - M_n$.

Recall in Chapter 5 how we first considered the concept of the standard error of a mean (σ_M). This was a mathematical way of expressing sampling error. Without going into the details of calculation, consider the concept of the *standard error of the difference between means*. Just as a single sample mean can be thought of as a statistic and its sampling error expressed in terms of the concept of standard error, so we can treat the difference between two sample means as a statistic and consider its standard error (that is, $\sigma_{\text{diff.}}$).

Figure 6.1 illustrates what a distribution of the sampling error of the mean difference might look like. Imagine that instead of just two samples, we had successively drawn and redrawn many pairs of samples. Each pair of samples has a mean difference (for example, $M_r - M_n$). If we took all

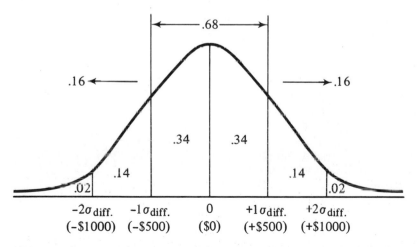

FIGURE 6.1   Sampling distribution of values of $M_r - M_n$ under the null hypothesis of $m_r = \mu_n$, and where $1\sigma_{\text{diff.}} = \$500$

of these values of mean differences, we could plot a distribution much like the one in Figure 6.1. Just as we saw how a distribution of sample means (Fig. 5.7) could have its own standard deviation, so can this distribution of mean differences. But in the present case, this standard deviation is called a *standard error of the mean difference*. In reality, we do not draw successive pairs of samples but calculate this value by use of a particular formula.[1]

Notice that the baseline of this distribution (Fig. 6.1) is divided into σ units but that these units represent standard errors of the mean difference ($\sigma_{\text{diff.}}$). We can also see how actual values of $M_r - M_n$ fit along this baseline because we would have calculated what $1\sigma_{\text{diff.}}$ is equal to in terms of actual measurement units. What this distribution represents is the different values of $M_r - M_n$ that we would expect because of sampling error, assuming that the curve is based on the null hypothesis, for instance, $\mu_r = \mu_n$ or $\mu_r - \mu_n = 0$. Put another way, if the two samples actually came from populations having the same mean, we could expect the kind of sampling error in values of $M_r - M_n$ as shown in Figure 6.1.

How, then, does probability enter the picture? Consider what we can interpret from the curve in Figure 6.1. Similar to the curves discussed in Chapter 5 (Fig. 5.4), its area represents probability. Thus, we can pick particular segments of the baseline of this distribution and interpret them in terms of the probability that some value will fall within them. What, for example, would be the probability of obtaining a value of $M_r - M_n = \$500$ or greater? According to the sampling error expected under the null hypothesis, the probability would be approximately .32 as shown in Figure 6.1. That is, the value of $500 could be either in the tail of the curve *below* $-\$500$, or $-1\sigma_{\text{diff.}}$, or in the tail *above* $+\$500$, or $+1\sigma_{\text{diff.}}$. The probability in each of these tails is equal to approximately .16, so if we consider that either tail might apply, the combined probability is .32. What would be the probability of occurrence under the null hypothesis of obtaining $M_r - M_n = \$1000$, or greater? Again, consulting the curve in Figure 5.1, we see that this value would fall in the tail below $-2\sigma_{\text{diff.}}$, or the tail above $+2\sigma_{\text{diff.}}$, and these combined areas would be interpreted as a probability of .04.

What we have seen, then, is the kind of a distribution that underlies how the probability of obtaining values of $M_r - M_n$ could be estimated. Again, this distribution represents the sampling error that would be expected under the null hypothesis. Hence, probabilities interpreted from it are the probabilities that we use when considering the hypothesis that $\mu_r - \mu_n = 0$ and that any $M_r - M_n$ differences we have observed are due solely to chance (sampling error).

In conducting actual analyses, we do not generally plot such curves. We calculate the value of $\sigma_{\text{diff.}}$; then, along with a value of $M_r - M_n$, we

[1] For example, $\sigma_{\text{diff.}} = \sqrt{\sigma_{M_r^2} + \sigma_{M_n^2}}$.

consult a table that gives us the probability value that would be expected under an appropriate curve. Chapter 7 on the t test will provide an example of an actual calculation.

THE RESULTS

The final step of the strategy for hypothesis testing involves interpretation of statistical results and, based on these results, stating a conclusion for the study. Ideally, this conclusion will be reasoned directly from the statistical results and will fulfill whatever goal was defined in the statement of the problem.

Interpreting Statistical Results

We have already seen in Figure 6.1 how, for a given difference between two sample means, we can estimate the probability of this occurrence under the terms of the null hypothesis. Let us now look again at this same distribution of sampling error under the null hypothesis, but this time, we will consider how the rejection regions fit in.

In the preceding example of the distribution of the statistic $M_r - M_n$, we considered probability in terms of both tails of the curve. Because the research hypothesis ($\mu_r \neq \mu_n$) said nothing about the direction of the difference between the means of the two populations, we considered the likelihood that a particular observed value of $M_r - M_n$ would be on *either* side of the curve. In more practical terms, because our research hypothesis said simply that the two population means would be different, but did not state the direction of this difference, we might expect to find a value of $M_r - M_n$ in either tail of the curve. Suppose, then, that we had set $p < .05$ as the significance level; can we see this rejection region in terms of the distribution of $M_r - M_n$ under the terms of the null hypothesis? The answer to this question is shown in Figure 6.2. Here we have the same curve as presented in Figure 6.1, but this time, we are considering it only in terms of where the rejection regions fall.

Note first that the portion of the curve incorporating rejection areas totals .05, the level of probability that we have set for rejection of the null hypothesis. This .05 has been divided into areas of .025 in each tail because the hypothesis we are testing may involve a value of $M_r - M_n$ that could fall in either tail. What this curve tells us is that if $1\sigma_{diff.} = \$500$, then a value of $M_r - M_n$ of either $-\$980$ or $+\$980$, or greater, will be within the region set for rejection of the null hypothesis. This is called a *two-tailed test*.

> *Two-tailed test:* a nondirectional hypothesis test that incorporates rejection regions in both tails of the probability curve used for a given statistic.

Suppose, however, that the research hypothesis said something about direction—for example:

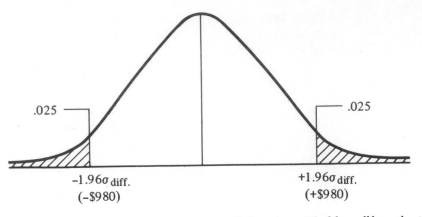

.025   .025

−1.96σ diff.   +1.96σ diff.
(−$980)   (+$980)

FIGURE 6.2   Rejection regions for a two-tailed test ($p < .05$) of the null hypothesis $\mu_r = \mu_n$ against the alternative hypothesis of $\mu_r \neq \mu_n$ and where $1\sigma_{\text{diff.}} = \$500$

$$\mu_r > \mu_n$$

The null hypothesis is then:

$$\mu_r \leq \mu_n$$

This kind of a null hypothesis—that is, one incorporating direction—need not be tested against both tails of the probability curve. If we were to obtain a value of $M_r - M_n$ that would serve as a basis for rejecting this null hypothesis, then we would expect it to be only a positive value, that is, a difference in the direction predicted by the research hypothesis. Because in this case we are testing the null hypothesis only in terms of what might be found in one tail of the curve, the rejection region is located as shown in Figure 6.3. If this rejection region has been set at $p < .05$, then we locate this entire region only in the tail of the curve that would be applicable in considering rejection of the null hypothesis. In this case, according to the distribution shown in Figure 6.3, a value of $M_r - M_n$ equal to $+ \$825$, or greater (and no negative values), would serve as grounds for rejecting the null hypothesis. In practical terms, we are simply testing the null hypothesis against values of $M_r - M_n$ that could occur in one direction only. Those that might occur in an opposite direction could not be used as a basis for rejecting a null hypothesis. Obviously, then, in setting up a directional research hypothesis, the researcher must be very confident beforehand that there is no reason to expect differences in an opposite direction. This is called a *one-tailed test*.

> *One-tailed test:* a directional hypothesis test that incorporates a rejection region in only one tail of the probability curve used for a given statistic.

.05

$+1.65\sigma_{\text{diff.}}$
$(+\$825)$

FIGURE 6.3   Rejection region for a one-tailed test $(p < .05)$ of the null hypothesis $\mu_r = \mu_n$ against the alternative hypothesis $\mu_r \neq \mu_n$ and where $1\sigma_{\text{diff.}} = \$500$

Again, a one-tailed test is never used (or should never be used) unless there is very good reason to make a directional prediction. This reason is not especially a function of wanting to make a directional prediction but is mainly based on the confidence that an outcome in the opposite direction will not be obtained. If a one-tailed test is made in the absence of such confidence, the researcher is assuming the hazard of never really testing statistically an outcome that would be opposite to the predicted direction.

The statistical result of a hypothesis-testing study is a probability value determined for the null hypothesis. If this probability value falls within the region set for rejection, the researcher then has a basis for rejecting the null hypothesis and accepting, in turn, the research hypothesis. As will be seen in Chapter 7, we never really plot such curves as shown in Figures 6.2 and 6.3, but instead, we use tables for interpreting statistical results.

Stating the Conclusion

Given a decision regarding the null hypothesis, a final step involves reporting the outcome of the hypothesis test. Let us consider how some of the alternative outcomes might be stated.

Suppose that the calculated value of probability were greater than that set as the significance level (rejection region). For example, if the rejection region had been set at $p < .05$, but the statistical results indicated that $p > .05$, in terms of the null hypothesis, we might say:

It is likely that the two samples represent populations with equal means,
(or)
There is no reason to believe that the two samples represent different populations in terms of their means,
(or)

The observed difference between the two sample means was not statistically significant.

In terms of the research hypothesis, when the null hypothesis has not been rejected, we might write something such as:

There was no evidence to support the research hypothesis that. . . ,

(or)

There was no reason to believe that. . . .

On the other hand, if the outcome were such that $p < .05$, the null hypothesis would be rejected. If we wished to say something about the null hypothesis, it might be:

It is likely that the two samples represent populations with different means,

(or)

There is reason to believe that the two samples represent different populations in terms of their means,

(or)

The observed difference between the two sample means was statistically significant.

The conclusion for the research hypothesis might be phrased something like:

There was evidence to support the research hypothesis that. . . ,

(or)

Results supported the research hypothesis that. . . .

ERRORS IN HYPOTHESIS TESTS

Thus far, you should be able to see that a good share of the logic in hypothesis testing involves nothing more than setting up a decision procedure for evaluating a research hypothesis. In discussing the logic, we have put the stress on which circumstances lead to support of a research hypothesis. A vital part of this logic can also be seen in terms of the kinds of *error* that we attempt to guard against in this decision procedure.

Type I and Type II Errors

The two main kinds of errors considered in the decision procedure for hypothesis testing are:

Type I error: rejecting a null hypothesis when it should have been the acceptable alternative.

Type II error: accepting a null hypothesis when it should have been the rejected alternative.

To aid us in gaining a conceptual understanding of these two types of error, let us return to the probability distribution of different values of $M_r - M_n$ expected under the null hypothesis. The curve in Figure 6.4

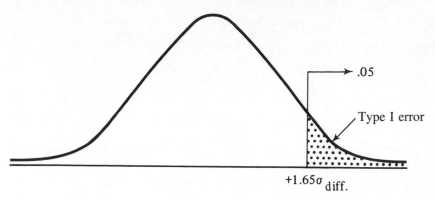

FIGURE 6.4   Rejection region for a one-tailed test ($p < .05$) of the null hypothesis, where the area (dotted) of the rejection region equals the probability of Type I error

is what we would expect if the null hypothesis were true. In terms of this curve, what we have set aside as a rejection region of $p < .05$ can also be seen as the probability of committing Type I error. Why is this the case? For one thing, the distribution in Figure 6.4 represents what we would expect in terms of the null hypothesis. When considering Type I error, we assume for the moment that this curve is the true or acceptable alternative as compared with the research hypothesis. What we are showing in the rejection region is the probability of rejecting this distribution as the acceptable alternative, or the probability of Type I error. Put into simpler terms, whatever level of probability we set as a basis for rejecting the null hypothesis is also the probability of our committing a Type I error.

Type II error is a little more difficult to conceptualize. Note first, as indicated in its definition, Type II error implies that some $M_r - M_n$ distribution other than that for the null hypothesis is the acceptable alternative. To conceptualize an example of this other distribution, let us assume that $\mu_r - \mu_n$ is really equal to $1500. That is, the two population means are indeed $1500 apart. We shall then compare these two distributions—that is, the null and the true alternatives—on the same graph, with a common baseline, in Figure 6.5. Given these two distributions, suppose that we set the significance level at $p < .05$. The area under the null hypothesis distribution, but to the right of the significance level, is the probability of Type I error. By contrast, the area under the distribution of the true alternative, but to the left of the significance level, is the probability of Type II error.

If the curves in Figure 6.5 tend to confuse you, try to remember the following. Type I error is where the null hypothesis was really true (thus, we consider area in the curve under the null hypothesis) but where we

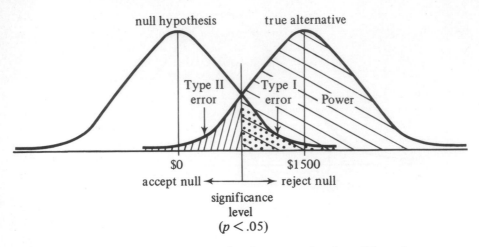

FIGURE 6.5   Comparison of $M_r - M_n$ distributions under the null hypothesis
and under a true alternative

have rejected it (thus, the area to the right of the significance level). By
contrast, Type II error is where the alternative hypothesis was really true
(thus, we consider area under the curve of the true alternative) but where
we have accepted the null hypothesis instead (thus, the area to the left
of the significance level).

With all other factors being equal, consider what the consequences are
of changing the significance level. Suppose, for example, that we set it
as $p < .01$. This would be shown by moving the significance level to the
right in Figure 6.5, thus decreasing the probability of Type I error but
increasing the probability of Type II error. The converse is also true.
Suppose that the significance level were set at $p < .10$. This would have
the effect of moving the significance level to the left in Figure 6.5, thus
increasing the probability of Type I error but decreasing the probability
of Type II error.

Power

There is an additional main factor that separately affects Type II error,
however. This is *statistical power*, a probability that is also shown in Figure
6.5.

Power: the probability of rejecting a null hypothesis that is, in fact, false.

Because power is based on the idea of the null hypothesis being false, we
are interested in some area within the distribution of the true alternative.
Also, because power is based on our acceptance of this alternative distribution, then we are considering all of the area to the right of the
significance level shown in Figure 6.5. One way of considering this area
is that it is equal to 1—the probability of Type II error. In other words,

all other factors being equal, if we increase the power of a test, we are in effect changing the picture shown in Figure 6.5, so that the amount of Type II error is accordingly reduced. All this would be independent of any effect on the probability of Type I error.

What we want, then, in a statistical test is power—the probability of rejecting a null hypothesis that is, in fact, not the acceptable alternative. If we were somehow to diminish power, we would be more prone to Type II error, that is, the probability of accepting the null hypothesis when it should be rejected. How do we get power? Suffice it to say that we gain power from certain factors in the design of the experiment (such as sample size) and from the kind of statistical models that we use for our calculations for the distributions.

Given a situation where a particular power is attained, the setting of the significance level is then a compromise between Type I and Type II error. That is, if we increase the probability of Type I error, we decrease the probability of Type II error and vice versa. Usually, this decision is based on the kind of error that the researcher is willing to tolerate. Suppose that we were engaged in so-called pilot research. We are more or less trying out an experimental design to see if some gross effects are observed. Here we might set the significance level at .10 or .20 because we are not greatly worried about the consequences of rejecting a true null hypothesis. We are simply looking for grounds to conduct a further, more rigorous, experiment. On the other hand, suppose that we could not tolerate Type I error to any great degree. Perhaps the experiment tests a prediction that we want to be highly confident about when the null hypothesis is rejected. Thus, we might set the significance level at .01 or even .001.

A NOTE ON FAILURE TO REJECT THE NULL HYPOTHESES

If a study results in failure to reject a null hypothesis, the researcher has not really "proved" a null hypothesis, but has failed to find support for the research hypothesis. It is not unusual to find studies with negative outcomes where the researcher has placed a great deal of stock in "acceptance" of null hypotheses. Such interpretations, strictly speaking, are in error because the logic of a research design incorporates the testing of some alternative (research hypothesis) against the status quo (null hypothesis). Although failure to find support for the alternative does leave one with the status quo, it *does not* rule out other possible alternatives. Put into practical terms, be skeptical of interpretations of unrejected null hypotheses.

SUMMARY

Hypotheses can concern predictions of relationship or differences. In the present chapter, the examples center upon predictions of differences. When concerned with such a prediction, we state a research hypothesis. Alternatively, for statistical purposes, we state a *null hypothesis*, which says that whatever difference we have observed between samples is due to sampling error. Sampling statistics provide us with a basis for estimating the probability that some observed difference between samples would be expected upon the terms of the null hypothesis—due to sampling error. If this obtained value of probability is equal to or less than some value preset as a *rejection region* (*significance level*), we reject the null hypothesis in favor of its logical alternative, the research hypothesis.

In stating a problem in hypothesis form, the research hypothesis must imply an unambiguous null hypothesis in terms of the population characteristics to be compared. Along with setting up these hypotheses, a significance level is also set. Sampling statistics then provide a basis for estimating the probability value that will eventually be compared with the significance level. Given this comparison, a decision is made with regard to the null hypothesis, and this in turn serves as a basis for a decision about the research hypothesis.

Much of the practical logic in hypothesis testing can be seen as a decision procedure. In considering this procedure, we often center upon the kinds and degrees of error that are involved. Two main types of error are *Type I*, which is the probability of rejecting a null hypothesis that should have been the acceptable alternative, and *Type II* which is the probability of accepting a null hypothesis that should have been the rejected alternative. Finally, the *power* of a test is seen as the probability of rejecting a null hypothesis that is the unacceptable alternative.

SUPPLEMENTARY READINGS

Most basic statistics texts have introductory chapters on hypothesis testing; here are a few such texts:

Hays, William L., *Statistics*. 3rd ed. New York: Holt, Rinehart and Winston, 1981. Chapter 7 is a thorough coverage of hypothesis testing.

Howell, David C., *Fundamental Statistics for the Behavioral Sciences*. Boston: PWS-Kent, 1989. Chapters 11 through 14 have many examples.

Kidder, Louise H., *Selltiz, Wrightsman and Cook's Research Methods in Social Relations*. 4th ed. New York: Holt, Rinehart and Winston, 1981. See Chapter 2.

PART FOUR

ANALYSES OF DIFFERENCES

7

The *t* Test

In the previous chapter, we saw how statistical logic is brought to bear in testing statements about population differences. For example, a research hypothesis stating that two population means differ will imply rejection of the null hypothesis. Under the terms of the null hypothesis, any differences that exist between the sample means of these two populations are due to sampling error. Using a statistical method called a t test, we can calculate the probability of whether a particular difference between sample means would be expected under the terms of the null hypothesis, that is, attributed to sampling error. If a particular probability is equal to, or less than, a level of probability set as a criterion for rejecting the null hypothesis, we reject the null hypothesis and accept its logical alternative, the research hypothesis, and conclude that the sample means reflect different population means.

In this chapter and in three subsequent chapters, we shall review in some detail the various statistical procedures used in tests of differences. These models each provide a basis for reasoning statistically from sample distributions in order to gain an estimate of the probability that whatever difference was observed between or among samples is due to sampling error. The present chapter centers upon the *t* test and how it is used for testing the difference between two population means, based on the observed difference between two sample means and their distributions. Although this is a common application of the *t* test, the reader should be aware that the *t* test itself has several other uses.

A SAMPLE STUDY

To illustrate how the *t* test of means fits into a research plan, we shall first consider a sample study. The data have been deliberately simulated in skeletal form for ease of understanding.

Problem

The experimenter is interested in assessing whether children's perception of aggressive behavior in a television cartoon will affect aggressive tendencies in their behavior. One theory reasons that children will imitate the aggression that they see in the cartoons, thus viewing will lead to an increase in children's tendencies toward aggressive behavior. A counter theory poses that any aggressive tendencies children might have are vicariously released by seeing the cartoon; hence, behavior will be markedly less aggressive than usual after viewing such a cartoon. Considering both of these theories, the researcher predicts that viewing aggression in a cartoon will have some effect on children's aggressive tendencies, but there is no prediction whether this will be an effect of increase or decrease in these tendencies. In more specific terms, the research hypothesis is that the mean ratings of the amount of aggressive tendencies of children who have seen an aggressive cartoon (Group I) will be different from the mean ratings of children who have not seen such a cartoon (Group II); that is:

$$R.H.: \mu_I \neq \mu_{II}$$

The implied null hypothesis is:

$$\mu_I = \mu_{II}$$

The researcher sets $p < .05$ (two-tailed) as the significance level.

Method

From a kindergarten class, 10 children are randomly selected to serve as subjects (Ss) in an experimental design.[1] Of these, five are randomly selected to serve in a group that sees a 10-minute televised cartoon in which the central character's behavior is predominantly aggressive. The remaining five children see an alternative cartoon under exactly the same conditions, except that there is no aggressive behavior depicted.

After exposure to the cartoons, both groups of Ss are placed in a playroom where their behavior is observed through a two-way mirror and rated by several judges. The ratings are in a numerical form for which the experimenter can assume the power of an interval scale, and the distributions presumably fit the normal curve model. Each child is assigned a value that represents the mean of the judges' ratings of aggressiveness. Given the measures of the two samples, the means are calculated:

$$M_I = 57 \qquad M_{II} = 52$$

The sampling statistic t will be used to test the null hypothesis against the alternative research hypothesis. A value of t is calculated according to an appropriate formula (as in Table 7.1). Values of t themselves have a sampling distribution that is used as a basis for estimating the probability that a particular value of t would be expected under the terms of the null hypothesis. In other words, under the null hypothesis, what is the probability that the value of t could occur because of sampling error? In this particular case (calculations in Table 7.1), $t = 2.50$. The probability of obtaining this value as a function of sampling error is $.02 < p < .05$, according to an appropriate table of the sampling distribution of t (Table

---

[1] Again, the size of sample is intentionally small for purposes of illustration.

Table 7.1   **Example of the Calculation of *t* (Skeletal Data)**

Grouped Data		General Formula
I	II	

$$t = \frac{M_I - M_{II}}{\sigma_{\text{diff.}}}$$

	I	II
	61	50
	53	48
	59	56
	55	54
	57	52
M	57	52
n	5	5

Specific Formula

$$t = \frac{M_I - M_{II}}{\sqrt{\left(\dfrac{\Sigma d_I^{\,2} + \Sigma d_{II}^{\,2}}{n_I + n_{II} - 2}\right)\left(\dfrac{n_I + n_{II}}{n_I \cdot n_{II}}\right)}}$$

Deviations from
Samples Means

d_I	d_{II}
4	−2
−4	−4
2	4
−2	2
0	0

$$= \frac{5}{\sqrt{\left(\dfrac{40 + 40}{5 + 5 - 2}\right)\left(\dfrac{5 + 5}{5 \cdot 5}\right)}}$$

$$t = \frac{5}{2} = 2.50$$

With $d.f. = 8$, $.02 < p < .05$

Deviations Squared

$d_I^{\,2}$	$d_{II}^{\,2}$
16	4
16	16
4	16
4	4
0	0
Σd^2 40	40

7.2).[2] A practical way to think about sampling error here is that if this comparison were repeated 100 times, we would expect that a *t* of 2.50 would occur between two and five times as a function of sampling error.

Results

Because the probability value meets the criterion for significance (that is, $p < .05$), the null hypothesis is rejected in favor of the research hypothesis. That is, $\mu_I \neq \mu_{II}$ is taken as the acceptable alternative. With this support, the researcher concludes that the different cartoons have caused the two groups of children, who were formerly sampled from the same population, now to represent different populations in terms of the measurement taken. The main conclusion is stated as:

[2] $.02 < p < .05$ is read as "probability greater than .02 but less than .05."

Table 7.2   Example of a Table of t (See p. 206 for a more standard, complete, version.)

d.f.	Probability				
	.20	.10	.05	.02	.01
1	3.08	6.31	12.71	31.82	63.66
2	1.89	2.92	4.30	6.97	9.93
. . .	. . .	. . .	. . .	. . .	. . .
8	1.40	1.86	2.31	2.90	3.36
. . .	. . .	. . .	. . .	. . .	. . .
18	1.33	1.73	2.10	2.55	2.88
. . .	. . .	. . .	. . .	. . .	. . .
Infinity	1.28	1.65	1.96	2.33	2.58

There is evidence to support the prediction that children viewing aggressive behavior in a televised cartoon will have a different tendency to engage in aggressive play than children who see a nonaggressive cartoon.

THE LOGIC UNDERLYING t

Given this sample experiment, just what has been involved in the process of reasoning with the t test? We can answer this question from two standpoints—what goes into the calculation of t and how values of t are interpreted.

Calculation of t

The logic in calculating t is similar to our discussion of the hypothesis test of difference presented in Chapter 6. As a test of the significance of difference between means, the general formula for t is given as follows:

$$t = \frac{M_I - M_{II}}{\sigma_{diff.}}$$

where M_I and M_{II} are the sample means and $\sigma_{diff.}$ is the standard error of the difference between the means. In the specific formula for calculating t, the numerator $(M_I - M_{II})$ is almost always the same.[3] It is the calculation of the denominator $(\sigma_{diff.})$ that varies in different formulas.

The concept of $\sigma_{diff.}$ is the same as discussed in Chapter 6. Again, imagine a distribution formed by drawing successive pairs of samples and plotting the $M_I - M_{II}$ value in each case. The standard deviation of this distribution is called the standard error of the mean difference, or simply, $\sigma_{diff.}$. For the t test, we calculate $\sigma_{diff.}$ in somewhat different ways, depending on

---

[3] Strictly speaking, the numerator reads "$M_I - M_{II}$" if the null hypothesis is $\mu_I = \mu_{II}$. Otherwise, the numerator reads "$(M_I - M_{II}) - (\mu_I - \mu_{II})$."

the sizes of the samples, whether the variances of the two samples are approximately the same, and whether the two sets of measures are somehow related.[4] Table 7.1 shows how t would be calculated for the sample experiment. Note in the denominator of the formula that the standard error of the mean difference ($\sigma_{\text{diff.}}$) is estimated from the variance observed in the two samples. The two main ingredients in this case are the sum of squared deviations about the two means (that is, Σd^2) and the sample sizes (n). All formulas for the calculation of t accommodate sample variability in one way or another.[5] Just as our calculation of standard error of the mean (σ_M) is affected by sample variance, so is our calculation of the standard error of the difference between means ($\sigma_{\text{diff.}}$).

Interpretation of *t*

In Chapter 6, the difference between means and the $\sigma_{\text{diff.}}$ was interpreted in terms of the normal probability curve. This same curve can be used for interpreting t values when the sample sizes are relatively large, say, 50 to 60 cases each. But as sample sizes grow smaller, the sampling distribution of t departs from normality; its center becomes increasingly peaked, and its tails become thicker. These changes in shape have consequences when determining values of probability for different values of t. Rather than try to depict all different shapes of the sampling distribution, it is more convenient to present it in tabular form. When interpreting a particular value of t for given sample sizes, the researcher consults a tabular form of a t distribution, such as shown in Table 7.2.

Notice that to enter a table of t (Table 7.2), it is necessary to consider what are called *degrees of freedom* (*d.f.*). In this table, degrees of freedom reflect sample sizes.[6] When two independent samples are being considered (as in the example experiment), *d.f.* are equal to the sum of the two sample sizes, minus the value 2 (that is, $n_I + n_{II} - 2$).[7] The different rows in the t table, each showing a different value of *d.f.*, accommodate the changes in the sampling distribution of t as the sizes of samples may vary. Given the row corresponding to the *d.f.*, for a particular comparison, the researcher then finds where the calculated value of t falls in terms of prob-

---

[4] "Related" refers to *correlation* between two sets of measures, a topic treated in Chapter 11. Correlated measures would be involved if the two sets of measures were on the same Ss or else if the two sets of Ss were somehow matched on traits relevant to the measure.

[5] For relatively large independent samples, the estimate of $\sigma_{\text{diff.}}$ is equal to $\sqrt{\sigma_{M_I}^2 + \sigma_{M_{II}}^2}$.

[6] Degrees of freedom enter into many different statistical procedures, as will be soon in subsequent chapters. The basic concept is not far removed from what the name implies. For example, given an array of scores of the size N and with a given mean, all scores except one could be varied at will and still preserve the same mean, so long as the one score compensated for the others in order to yield that mean. In this case, *d.f.* is equal to $N - 1$. Given two subgroups of the combined size N, the means of the two subgroups could be preserved while varying all save one score in each subgroup; hence, *d.f.* here is equal to $N - 2$, as in the present example.

[7] For correlated measures, *d.f.* is equal to the number of *pairs* of scores minus the value 1.

ability values. In the example experiment, for $d.f. = 8$, a t of 2.50 falls
between the .05 and .02 columns. This means that probability will be
reported as $.02 < p < .05$. In this same row, a value of t exceeding 3.36
would be reported as $p < .01$. But if t were less than 1.40, probability
would be designated as $p > .20$.

Although Table 7.2 is labeled for two-tailed tests, it can be adapted to
one-tailed tests by halving the probability values at the top of each column.
Thus, if our t were equal to 2.50 and if we were using a one-tailed test
with 8 $d.f.$, probability would be $.01 < p < .025$. (We are still between
the columns labeled .02 and .05, but we halve these probability values
to .01 and .025, respectively.)

By examining Table 7.2 further, we can see the consequences that
varying sample sizes will have on the interpretation of t. Note, for example,
that with 8 $d.f.$, t must be 2.31 or greater to be associated with $p < .05$.
But if the samples were such that $d.f. = 2$, then a larger value of t (4.30)
would be necessary to meet the $p < .05$ criterion. When considering solely
the interpretation of the sampling distribution of t, as sample sizes become
smaller and all other factors are equal, a larger value of t is necessary to
gain a particular value of probability. On the other hand, as sample sizes
become larger, a smaller value of t would suffice. But as the sampling
distribution of t approaches normality, further increases in sample size
no longer have this particular effect. When sample sizes approach 100 or
greater, many researchers use the normal curve model for interpreting
t. This would be the same as entering Table 7.2 for $d.f.$ equal to infinity,
which is the same as the normal curve model (Chapter 5).

The sampling distribution of t tells us the probability that, in accord
with the null hypothesis, we could expect a particular value of t. Put
another way, the sampling distribution tells us what sampling error to
expect in values of t.

SOME GENERAL CONSIDERATIONS OF *t*

Given what has been said about the t test, what are some characteristics
of t that are useful to remember? For one thing, as logically suggested
by the research hypothesis, as the difference between the sample means
$(M_I - M_{II})$ is increased, and all other factors are equal, the value of t will
increase. And as indicated by the sampling distribution of t, as t grows
larger, associated values of probability (p) grow smaller. As can be seen
in the calculation (Table 7.1) of the denominator of the t formula, the
greater the variation is among the scores of each sample, and all other
factors being equal, the greater will be the estimate of $\sigma_{\text{diff.}}$ and, hence,
the smaller the value of t. In other words, the more variability there is
among scores in a sample, the more difficult it will be to gain a t for a
particular significance level.

One logical question sometimes comes to mind when people think of likely extensions of studies using *t* tests: What if we want to compare more than two groups on a dependent variable? The answer is that we use a technique called *analysis of variance*, the topic of the next two chapters.

Finally, along another line, what if we want to compare two groups in terms of more than one dependent variable? There are several alternatives, the most typical of which is *multiple discriminant analysis*, the topic of Chapter 16.

SUMMARY

The *t* test is a statistical model that can be used for testing the significance of difference between the means of two populations, based on the means and distributions of two samples. The logic of *t* is similar to the test of the difference between population means as discussed in Chapter 6; that is, it is a ratio between the sample mean difference and the standard error of that difference. The unique features of *t* are found in the various methods for estimating the standard error of the mean difference and in the fact that the sampling distribution of *t* departs from normality when small samples are used. Given a calculated value of *t*, this value is interpreted for its probability of occurrence in testing a null hypothesis against an alternative research hypothesis. If this probability value is equal to or less than the level set for significance, the null hypothesis is rejected in favor of the research hypothesis.

SUPPLEMENTARY READINGS

Rosenthal, Robert, and Ralph L. Rosnow, *Essentials of Behavioral Research: Method and Data Analysis*. New York: McGraw-Hill, 1984. See Chapter 18.

SPSS, Inc., *SPSS-X User's Guide*. 3rd ed. Chicago: SPSS, 1988. Chapter 49 describes a program for conducting *t* tests.

8

Single-Factor
Analysis of Variance

Whereas the t *test is usually used in testing the difference between two population means based on differences between sample means, analysis of variance is used when the research hypothesis incorporates two or more population means and tests of differences among their respective sample means.*

SINGLE-FACTOR MODEL

What does *single factor* mean? Consider first that the term *factor* can mean the same thing as *independent variable*. In our examples in previous chapters, we have seen an independent variable as involving a distinction in treatments administered by the researcher to the different groups of subjects involved in an experiment. Thus, in the preceding chapter, the kind of cartoon that the children saw was the independent variable. We would describe this independent variable as having two levels: viewing a cartoon showing no aggression and viewing a cartoon showing aggression. In that example, the research was concerned only with the two postexposure sample means and whether they reflected the same or different population means. Suppose, however, that the experiment had three levels of the cartoon factor. This would involve a statistical model that could accommodate a test of more than two means. As described in the subsequent example, single-factor analysis of variance would be the likely statistical model in this case.

A SAMPLE STUDY

For the present purposes, we shall extend the hypothetical experiment described in the preceding chapter. Again, the samples are intentionally small and simulated for ease of presentation.

Problem

Given the finding that viewing aggressive actions in a televised cartoon will affect aggressive tendencies in a child's own behavior, a further question is raised. Will there be such an effect if the aggressive behavior of the cartoon character is punished as a part of the cartoon drama? Seemingly, if children see that aggression leads to punishment, then perhaps they will avoid a tendency in their own behavior or at least lessen it.

A second experiment is designed to study this problem. This time, three groups of children are incorporated into the study. Again, the children's behavior will be assessed after exposure to a cartoon. But this time each group will be exposed to a different cartoon: one that portrays no aggression, aggression that goes unpunished, or aggression that is punished. The prediction is that after seeing the different cartoons, the groups will have different mean tendencies to engage in aggressive play.

In mathematical form, the research hypothesis is the overall statement:

$$\mu_I \neq \mu_{II} \neq \mu_{III}$$

The implied null hypothesis is the overall statement:

$$\mu_I = \mu_{II} = \mu_{III}$$

The level set for rejection of the null hypothesis is $p < .05$.

Method

From a kindergarten class, 15 children are randomly selected to serve as Ss in the experiment. In turn, they are randomly assigned to the following groups, each with a respective treatment.

Group I ($n = 5$) sees a 10-minute televised cartoon in which the central character shows no aggressive behavior.

Group II ($n = 5$) sees a 10-minute televised cartoon in which the central character engages in aggressive behavior, and the cartoon does not include any punishment of this behavior.

Group III ($n = 5$) sees a 10-minute televised cartoon in which the central character engages in aggressive behavior, but this behavior is subsequently punished.

The independent variable, then, comprises the differences in the above three treatments, or three levels of the cartoon factor. The dependent variable is the Ss' tendency to engage in aggressive play. As in the previous example (Chapter 7), the measure is a numerical rating based on observations by judges.

Means for the three samples are:

$$M_I = 51 \qquad M_{II} = 58 \qquad M_{III} = 53$$

The researcher uses analysis of variance as a method for making a probability statement about the null hypothesis. The calculations yield a statistical value called F. This value can be interpreted in a sampling distribution to determine its probability associated with occurrence under the null hypothesis. In this study (calculations in Table 8.1), F is equal to 16.25. The associated probability value (Table 8.2) is $p < .01$. That is, under the hypothesis of sampling error (null hypothesis), we would expect to find the present differences among sample means in fewer than one out of 100 cases of random sampling.

Table 8.1   **Example of a Single-Factor Analysis of Variance (Skeletal Data)**

Total SS (Sum of Squares)	General Formula

Score deviations (d_g) from grand mean (54)

$$F = \frac{MS \text{ between}}{MS \text{ within}}$$

Group I	d_g	d_g^2
49	−5	25
52	−2	4
52	−2	4
53	−1	1
49	−5	25

Summary Table

Source	SS	d.f.	MS	F
Between	130	2	65	16.25
Within	48	12	4	
Total	178			

Group II		
56	2	4
57	3	9
57	3	9
60	6	36
60	6	36

Within SS

Source deviations (d_w) from group means:

I	II	III
−2	−2	1
1	−1	−1
1	−1	3
2	2	−3
−2	2	0

Group III		
54	0	0
52	−2	4
56	2	4
50	−4	16
53	−1	1

Deviations squared (d_w^2):

I	II	III
4	4	1
1	1	1
1	1	9
4	4	9
4	4	0

Total SS = $178 = (\Sigma d_g^2)$

Sums   14   14   20

Within SS = $14 + 14 + 20 = 48 \ (\Sigma d_w^2)$

Between SS

	I	II	III
Group means .... (M)	51	58	53
Group deviation from grand mean .... (d_b)	−3	4	−1
Square deviation .... (d_b^2)	9	16	1
Group n times d_b^2 .... (nd_b^2)	45	80	5

Between SS = $45 + 80 + 5 = 130 \ (\Sigma nd_b^2)$

Results

Because the obtained probability value meets the criterion for statistical significance, the null hypothesis is rejected in favor of the research hypothesis. The researcher concludes:

> There is evidence that a child's tendency to engage in aggressive behavior will be differentially affected under conditions where a televised cartoon portrays no aggression, unpunished aggression, or punished aggression.

Table 8.2   **Example of a Table of F** (*See pp. 207–208 for a complete version.*)

d.f. Denominator	d.f. Numerator				
	1	2	. . .	8	. . . 12
6	5.99	5.14	. . .	4.15	. . . 4.00
	13.70	**10.90**	**. . .**	**8.10**	**. . . 7.72**
. . .	. . .	. . .	. . .	. . .	. . . . . .
12	4.75	3.89	. . .	2.85	. . . 2.69
	9.33	**6.93**	**. . .**	**4.50**	**. . . 4.16**
. . .	. . .	. . .	. . .	. . .	. . . . . .
16	4.49	3.63	. . .	2.59	. . . 2.42
	8.53	**6.23**	**. . .**	**3.89**	**. . . 3.55**
. . .	. . .	. . .	. . .	. . .	. . . . . .
120	3.92	3.07	. . .	2.02	. . . 1.83
	6.85	**4.79**	**. . .**	**2.66**	**. . . 2.34**
. . .	. . .	. . .	. . .	. . .	. . . . . .
Infinity	3.84	3.00	. . .	1.94	. . . 1.75
	6.63	**4.61**	**. . .**	**2.51**	**. . . 2.18**

THE LOGIC UNDERLYING *F*

A first point of special note is that the hypothesis to be tested is considered as an overall statement. That is, analysis of variance will tell us only if there is significant variation among the means in that total statement; it will not tell us about the comparison of individual means, that is, μ_I versus μ_{III}, and so on. In other words, analysis of variance centers upon the question of whether the three samples represent the same population in terms of their means.

The Calculation of *F*

In the analysis of the sample experiment, we could see how the mean scores varied for each of the three groups. Logically, the more that these groups differ from each other in terms of the trait being measured, the more variability we would expect among their means. We can depict such variability statistically in terms of how the group means vary about a *grand mean*, that is, a mean of all of the groups combined. In fact, we can calculate the groups' variance about the grand mean. In analysis of variance, this is called the *between-groups* variance. More specifically, the more difference that there is among the group means, the greater would be the value of between-groups variance. What we need next is some way to interpret this amount of variance—namely, what is the probability of it occurring under the terms of the null hypothesis (that is, due to sampling error)?

Suppose that we assumed that all three groups came from the same population. If this were so, how much variability among individual scores would we expect in this population? In analysis of variance, this variability is actually estimated. It is based on the variability that we observed within each of the groups—aptly called *within-groups* variance.

The central point in analysis of variance is that if there are no differences among the groups, then the between-groups variance and the within-groups variance will be approximately equal. In other words, the more that the between-groups variance exceeds the within-groups variance, the greater is the probability that the groups represent different populations.

With the foregoing in mind, consider the definition of F:

$$F = \frac{\text{variance between groups}}{\text{variance within groups}}$$

If the null hypothesis were true and if there were no sampling error, we would expect F to equal 1.0. But in reality, we face the prospect of sampling error; therefore, we again use a sampling distribution to gain a probability statement. Under the assumption of the null hypothesis, the sampling distribution of F tells us the probability that we can expect the between-groups variance to be so many times greater than the within-groups variance. As with other hypothesis-testing statistics, if this probability meets a preset criterion level, we shall consider it as grounds for rejection of the null hypothesis.

Table 8.1 illustrates how F could be calculated for the sample experiment. Notice that most of the calculations refer to sums of squares (SS). The total SS reflects squared deviations of all scores about the grand mean, without any regard to group assignment. This total amount of squared deviation is partitioned into two components: squared deviations between the groups and the grand mean, and squared deviations of the scores within groups about their own means. These component squared deviations are reduced to mean squares (MS), which are actual calculations of variance. The degrees of freedom used to make these reductions represent the number of groups minus one (between $d.f.$), and the number of Ss minus one in each group times the number of groups (within $d.f.$).

Usually, the components of variation used to calculate F are displayed in a summary table similar to the one shown in Table 8.1.

Interpretation of F

As mentioned above, we use a sampling distribution of F to determine the probability that a particular value of F could occur under the terms of the null hypothesis. This sampling distribution changes quite radically according to the degrees of freedom in the numerator and denominator of the F ratio (in this case, reflecting the number of groups being compared and the size of the groups). An abbreviated version of the F sampling distribution is presented in Table 8.2. This table is entered at the intersection of a column and row representing the degrees of freedom for the numerator and denominator of the F ratio, respectively. In this particular example of an F table, the numbers in boldface and regular type represent the size of F necessary for probability of .01 and .05, respectively.

For the F of 16.25 in the sample experiment and with $d.f. = 2/12$, the probability value is less than .01. With $d.f. = 2/12$, an F of 3.89 would yield a probability of .05. We are, therefore, quite well below the level set for significance.

In considering the sample table of F (Table 8.2), notice that as degrees of freedom increases in the denominator of the F ratio, a smaller F is necessary to gain significance at a particular probability level. What this reflects is what one gains by having larger samples in the various experimental groups. Notice, however, that such an increase in degrees of freedom soon pays diminishing returns; for example, with $d.f. = 1/120$, an F of 3.92 is required for significance at the .05 level, but this required value of F only decreases to 3.84 with an infinite number of degrees of freedom in the denominator.

SOME GENERAL CONSIDERATIONS OF F

One useful point to remember about an F test is that the denominator represents our estimate of the variability among individuals in the population that we are studying. The numerator, by contrast, reflects whatever differences in variability are observed as a function of differences among the groups being compared. The ratio itself simply tells us by how many times the between-groups variability exceeds our estimate of the population variability. Finally, the sampling distribution of F gives us an estimate of the probability that the between-groups variance could be some multiple of the population variance. In other words, the F statistic is our basis for obtaining a probability statement referring to the null hypothesis. It is logical, then, that for a particular set of degrees of freedom, the larger the value of F, the smaller is the probability. That is, in terms of the null hypothesis, the larger the value of F, the less we would expect the null hypothesis to prevail In two-group designs, F and t are related ($F = t^2$); they are assessing the same variability.

THE PROBLEM OF COMPARING INDIVIDUAL MEANS

When considering the results of the sample study, all that was said reflected only significant variation among the three groups. The analysis of variance results said nothing about μ_I being different from μ_{II}, or μ_{II} being different from μ_{III}, and so on. We only knew that somewhere among the means there was a significant difference (or differences) leading to the significant F. By rejecting the null hypothesis, we are only taking the position that the overall statement of "$\mu_I = \mu_{II} = \mu_{III}$" is not acceptable.

Certainly, it is usually of research interest to know exactly whether each of these means is different from each other, and there are statistical procedures for making such comparisons. Often, such methods are called *follow-up* or *post hoc* tests. These labels reflect a critical question in making such tests: In conducting individual comparisons of the means, are we reasoning according to predictions made in designing the experiment, or are we simply looking after the fact to see if any significant differences show up? This is an important question because if we have planned all along to test particular hypotheses of individual mean differences, we should bring the most powerful statistical tests to bear in assessing these differences. We must take great care when we are making tests after the fact, that is, when we are letting the data suggest hypothesis tests. Most statisticians agree that if we are simply looking for any differences that might result, we should be cautious in what differences we call significant. Being cautious involves the use of more conservative tests, often those with *protection levels* that tend to lessen Type I error (that is, rejection of a null hypothesis that probably should not be rejected).

In practical terms, what the foregoing suggests is that we should take special care in selecting the particular type of test to use in making comparisons among individual means. There are many procedures available[1] for such tests, but it often becomes a relatively subjective decision to select the appropriate procedure to use. Depending on how conservative a particular procedure is, a difference between means that might be significant with one approach might not be significant with another. In more practical terms, the conclusions that a researcher draws concerning individual mean comparisons are only as valid as the decisions that were involved in planning and selecting such tests—whether or not they were planned beforehand—and the amount of protection deemed realistic in the tests themselves.

Despite the problems of individual mean comparisons, if a researcher chooses to interpret any differences whatsoever among individual means, the statistical basis for such interpretations is incomplete without application of the proper follow-up tests.

To illustrate how individual mean comparisons facilitate describing the outcome of a study, suppose that we consider further the results of the sample experiment. The overall F ratio provided a basis for deciding that there was a significant difference among the three treatment means (M_I = 51, no aggression; M_{II} = 58, unpunished aggression; M_{III} = 53, punished aggression). After finding this difference, we are interested in individual comparisons among the means. Which ones are significantly different from one another? Consider that a follow-up test offering a moderate amount of protection against Type I error is applied (calculations are not

---

[1] Some of the more frequently encountered ones are the *Tukey, Scheffé, Duncan,* and *Newman-Keuls* procedures. See the supplementary readings at the end of this chapter.

shown).[2] The rejection level for these tests is set at $p < .05$. One way of reporting the results is as follows:

$$51_a* \qquad 58_b \qquad 53_a$$

* (Means with common subscripts are not significantly [$p < .05$] different from one another.)

Here the follow-up tests indicate no difference between the group (I) who saw no aggression and the group who saw punished aggression (III), but both of these groups have means significantly different from the group who saw unpunished aggression (II). Notice that the researcher's concluding statement can now say much more than was reported earlier.

> There is evidence that a child's tendency to engage in aggressive behavior will be differentially affected under conditions where a televised cartoon portrays unpunished aggression compared with no aggression or punished aggression. A post hoc analysis of the present results indicated that the effects of no aggression and punished aggression were not significantly different from one another but that both were different from the condition where aggression went unpunished in the cartoon.

SOME FURTHER NOTES ON SINGLE-FACTOR MODELS

Although what has been discussed thus far should provide you with an idea of the general logic involved in a single-factor analysis of variance, this has been a cursory picture at best. Within the scope of the present treatment, a few additional notes can be provided to describe some further considerations in dealing with single-factor models. Many of these considerations also carry implications regarding the specific procedures for calculating and interpreting F ratios.

Random Compared with Fixed Factors

One central distinction made when considering the different levels of a factor (independent variable) is whether these levels represent discrete variables or whether they are points along some type of continuous scale. For example, the different cartoons used in the sample study represent discrete phenomena; as such, this variable is called a *fixed* factor. On the other hand, suppose that we had a series of cartoons that, on a predetermined basis, was taken as representing different degrees of aggressiveness. In this case, the levels of the factor might be taken as sampling points along a continuum of aggressiveness in cartoons. We would call this a *random* factor; its levels are considered a basis for "sampling" from a continuum of cartoon aggressiveness as an independent variable.

[2] D. B. Duncan, "Multiple Range and Multiple F Tests," *Biometrics*, XI (1955), 1–42.

Levels of a Factor

Obviously, one way of increasing the complexity of a single-factor model is to incorporate additional levels of the factor into the design. The main criterion for defining levels is primarily a compromise between theory and the requirements of design. Suppose, for example, that we were interested in the length of the cartoon as the central factor in stimulating aggressiveness. As a random variable, we might want levels corresponding to 5, 10, 15, 20, and 30 minutes of cartoon length. So long as the different conditions all represent different levels of the same independent variable, the model remains a single-factor design.

Repeated Measures

In the present sample experiment, different groups of Ss were tested in each of the three levels of the cartoon factor. It is sometimes the case, however, that the same group of Ss is tested across the different levels of a factor. This again is another consideration that takes into account theoretical as well as design problems. In cases such as the sample experiment, it is advisable that an S be exposed to only one level of a factor, lest his or her exposure to one level influence exposure to another level. By contrast, consider a study in which the independent variable is the length of time following exposure to aggression in a cartoon and the dependent variable is again the Ss' tendency to engage in aggressive play. In this situation, it is quite realistic to obtain successive measures on the same Ss; they might be taken, for example, at intervals of 30 minutes, 1 hour, 2 hours, and 4 hours. Such a design would involve *repeated* measures on the same subjects.

Analysis of Covariance

In most experiments, we take all possible precautions to ensure that prior to whatever experimental treatments are administered, we have good reason to believe that the Ss in different groups are equal in terms of the behavior that is to be assessed. As in the sample experiment, such equalization is usually provided by carefully adhering to random assignment. But what happens if we cannot be sure of such assignment? Consider, for example, that in studying kindergarten children we had no other choice than defining three groups as simply three different schoolroom classes. Perhaps the original assignment of the children to these classes was on a more-or-less random basis, but we cannot be sure that the three classes are equal in terms of the behavior we want to study. Sometimes it is possible to equalize such pretreatment differences by a statistical method. This method is called *analysis of covariance*.

Suppose that in the sample experiment, the researcher was faced with using three groups that were in fact different kindergarten classes. And perhaps one of these groups has slightly more aggressive children than the others; thus, there are group differences even before the experiment is conducted. The practical consequence of this situation is that if the experimenter finds a difference after the treatments, can it be attributed to the cartoons that were shown, or were the groups simply different to begin with? Use of analysis of covariance in this situation would proceed generally as follows.

Prior to exposing the children to the experimental cartoons, their aggressive tendencies are assessed by the judges. With such measures, it is possible to have some idea of pretreatment intergroup differences among the children in terms of the trait to be studied. We call these pretreatment ratings a *control* variable (or a *covariate*).

The experiment is conducted as previously described. The groups are exposed to different treatments and subsequently are rated in terms of aggressiveness in behavior. We consider the posttreatment mean scores in this case as the *criterion* variable.

What analysis of covariance provides is a method by which we can remove pretreatment variations (as measured by the control variable) from the posttreatment means (criterion variable) prior to testing the significance of the posttreatment differences among the groups. In more simple terms, analysis of covariance provides a basis for ruling out pretreatment differences when our interest is in testing posttreatment differences. The significance test for analysis of covariance uses an F ratio, and this is interpreted for probability in a manner similar to a straightforward analysis of variance.

Obviously, the control variable is of little use unless it is reasonably correlated with the dependent or criterion variable. Another problem is that once the criterion variable is "adjusted" for pretreatment differences, it no longer exactly represents the behavior of the experimental subjects in the given situation. In other words, an assumption must be added that the adjustment will not jeopardize the interpretation of these results as representing some population of interest to the researcher. Finally, it is always better to randomize the assignment of subjects to treatment groups than to depend on some subsequent type of adjustments.

Other Variations

If we want to extend our basic analysis of variance design to include more than one dependent variable, this is possible through the use of the more complex *multivariate analysis of variance* (not included in this text; see Kerlinger and Pedhazur, 1973). If we wish to assess the degree to which a group of variables can aid us in differentiating among individual or experimental treatments, *multiple discriminant analysis is used* (Chapter 16). Or as discussed in Chapters 11, 12, and 13, relationships among independent and dependent variables can be assessed by use of *correlation* or *regression* methods.

Usually, however, the most frequent question about extending analysis of variance involves the use of more than one independent variable in a design, the topic of the next chapter.

SUMMARY

Single-factor analysis of variance is a statistical model used for testing the significance of difference among two or more means when these means reflect the consequences of different levels of a single independent variable. The statistical logic of analysis of variance is incorporated in the *F* ratio, a ratio of *between-groups* variance to *within-groups* variance. Given a calculated value of *F*, this value is interpreted in a sampling distribution for its probability under the terms of the null hypothesis. If this probability value is equal to or less than the criterion set for statistical significance, the null hypothesis is rejected in favor of the research hypothesis.

Because analysis of variance tests only the overall hypothesis of differences among means, it is usually necessary to conduct subsequent tests between individual means, often called *follow-up* tests. Some special considerations of analysis of variance models include whether the independent variable is a *fixed* or *random* factor, the number of *levels* of the factor that are incorporated into the design, and whether *repeated measures* are involved. *Analysis of covariance* is a method by which known pretreatment intergroup variations can be mathematically removed from measures of posttreatment intergroup variations prior to conducting an *F* test.

SUPPLEMENTARY READINGS

Howell, David C., *Fundamental Statistics for the Behavioral Sciences*. Boston: PWS-Kent, 1989. See Chapter 16.

Kerlinger, Fred N., and E. I. Pedhazur, *Multiple Regression in Behavioral Research*. New York: Holt, Rinehart and Winston, 1973.

SAS Institute, Inc., *SAS User's Guide: Statistics*. 1982 ed. Cary, N.C.: SAS Institute, 1982. See the section on "Analysis of Variance" (pp. 113–255), especially pp. 128–130 for a one-way model.

Spence, Janet T., John W. Cotton, Benton J. Underwood, and Carl P. Duncan, *Elementary Statistics*. 4th ed. Englewood Cliffs, N.J.: Prentice Hall, 1983. See Chapter 12.

SPSS, Inc., *SPSS-X User's Guide*. 3rd ed. Chicago: SPSS, 1988. Chapter 39 describes a program for one-way analysis of variance.

Winer, B. J., *Statistical Principles in Experimental Design*. New York: McGraw-Hill, 1971. A classic text that includes an excellent introduction to analysis of variance.

9

Multiple-Factor Analysis of Variance

In Chapter 8, we saw how analysis of variance was applied to test the significance of difference among the means of three groups that varied in terms of a single factor. As an independent variable, this factor had three levels, each corresponding to the particular type of cartoon that was viewed by the children. We now turn to models that incorporate more than one factor.

MULTIPLE-FACTOR MODEL

Suppose that a study has more than one factor. What this means is that the design incorporates more than one independent variable, and each of these variables (or factors) can have two or more levels of its own. Usually, we call such approaches *factorial designs* and describe them in terms of the number of factors and the number of levels each factor has. Thus, for example, a 2×2 design (usually read as "two-by-two") incorporates two factors, each having two levels, or a $4 \times 2 \times 2$ design has three factors, the first having four levels, the second having two levels, and the third having two levels. In a given design, factors and their levels define different subgroups in the experiment. Multiple-factor analysis of variance provides methods for testing whether different subgroups, or various combinations of subgroups, represent different populations in terms of what is being measured as the dependent variable.

A SAMPLE STUDY

For purposes of illustration, we shall again return to the type of study used as an example in the last two chapters. As before, the groups are intentionally small, and the data are simulated for ease of presentation.

Problem

It is generally agreed that kindergarten-aged boys have a greater tendency to engage in aggressive play than do girls of the same age. In considering the finding (sample studies in Chapters 7 and 8) that viewing unpunished aggression in a televised cartoon will affect a child's tendency to engage in aggressive play, the foregoing raises a question about the effects according to a child's sex: Will boys and girls be similarly affected by viewing a television cartoon showing unpunished aggression compared with a cartoon showing no aggression?

The above question suggests the need for a two-factor experiment. One factor will be the distinction in the type of cartoon viewed; its levels will correspond to whether the cartoon is one showing no aggression or one depicting unpunished aggression. The second factor will incorporate levels making the distinction between male and female children viewing the cartoons. Again, the dependent variable will be ratings of the children's behavior during a play period after viewing one or the other of the cartoons.

The experiment, then, has two independent variables, the combinations of which are shown in Table 9.1. This is a 2×2 model, or a design incorporating "sex $\times$ cartoons." One of the useful features of a factorial design is that it allows us to test a number of different hypotheses in a single study. Note the potentially different population means shown in Table 9.1. For one thing, we can assess the effects of the two independent variables, one at a time. These are called *main effects*. In other words, quite apart from the different cartoon levels, do the male (m) and female (f) children represent the same population in terms of their mean ratings? Or, apart from the sex of the child, do the children in the groups who saw the aggressive (A) cartoon represent a different population than the groups who saw the nonaggressive (N) version? Suppose that the two research hypotheses (R.H.) and their corresponding null hypotheses (N.H.) are:

$$
\begin{array}{ll}
\text{(R.H.)} - \text{sex)} & \mu_m \neq \mu_f \\
\text{(N.H.)} & \mu_m = \mu_f \\
\text{(R.H.} - \text{cartoon)} & \mu_N \neq \mu_A \\
\text{(N.H.)} & \mu_N = \mu_A
\end{array}
$$

We say that these hypotheses are tested by assessing the *main effects* of the two independent variables.

Neither of the main effects hypotheses, however, directly addresses itself to the original question; that is, will the two cartoons have differential

Table 9.1   Diagram of the Two-Factor Sample Experiment

		Cartoon		
		No aggression (N)	Aggression (A)	
	Male (m)	μ_{Nm}	μ_{Am}	μ_m ⎤ Sex
Sex				⎬ main effect
	Female (f)	μ_{Nf}	μ_{Af}	μ_f ⎦
		μ_N	μ_A	
		____ Cartoon main effect ____/		

effects on boys as compared to girls? The answer to this question requires that we somehow consider the effects of the two independent variables (cartoon, sex) in a combined form. Such a combined form in a factorial design is called an *interaction*. The nature of this interaction is the third hypothesis that can be tested in the present design.

Consider that our predicted interaction is that viewing an aggressive cartoon, as compared with a nonaggressive cartoon, will indeed have a different effect on boys as compared with girls. This prediction could be stated in the following mathematical form:

$$(\text{R.H.}) \qquad (\mu_{Nm} - \mu_{Am}) \neq (\mu_{Nf} - \mu_{Af})$$
$$(\text{N.H.}) \qquad (\mu_{Nm} - \mu_{Am}) = (\mu_{Nf} - \mu_{Af})$$

We say that the above hypothesis is tested by the interaction effects of the experiments.

Finally, suppose that the criterion set for rejection of each of the null hypotheses is $p < .05$.

Method

Using a kindergarten population as a pool of potential Ss for the experiment, two groups are initially selected according to sex. These groups constitute 10 males (m) and 10 females (f). Each of these groups of 10 is, in turn, randomly divided into two groups of five children each for the two different cartoon treatments (N, A). The four groups receive the following treatments:

Groups Am and Af see a 10-minute televised cartoon (A) in which the central character engages in aggressive behavior that goes unpunished.

Groups Nm and Nf see a 10-minute televised cartoon in which the central character shows no aggressive behavior (N).

As in the previous examples (Chapters 7 and 8), the dependent variable comprises judges' ratings of the children's tendencies to engage in aggressive behavior during a play period following the cartoon viewing.

When the data are analyzed, the means are as shown in Table 9.2.

Table 9.2   Means for the Sample Experiment

		Cartoon		
		No aggression (N)	Aggression (A)	
	Male (m)	$M_{Nm} = 51$	$M_{Am} = 59$	$M_m = 55$
Sex				
	Female (f)	$M_{Nf} = 45$	$M_{Af} = 49$	$M_f = 47$
		$M_N = 48$	$M_A = 54$	

Grand $M = 51$

Notice that this table of means shows all of the means of the various subgroups that will be considered in testing the main effects and interaction effects of the study. The researcher uses a multiple-factor analysis of variance model as a method for deriving probability values for the three null hypotheses. For each of these null hypotheses, an F ratio is calculated (Table 9.3), then each of these values of F is interpreted in an appropriate sampling distribution (Table 8.2) to determine its probability of occurrence under the terms of the null hypothesis. That is, the question is asked in each case what the probability is of obtaining a particular value of F, if F

Table 9.3   **Example of a Multiple-Factor Analysis of Variance (Skeletal Data)**

TOTAL SS (SUM OF SQUARES)
—where d_g equals deviation of scores about the grand mean (51).

Group Am	d_g	d_g^2	Group Nm	d_g	d_g^2
57	6	36	49	-2	4
58	7	49	52	1	1
58	7	49	52	1	1
61	10	100	53	2	4
61	10	$\underline{100}$	49	-2	$\underline{4}$
		$\Sigma = 334$			$\Sigma = 14$

Group Af			Group Nf		
47	-4	16	43	-8	64
50	-1	1	46	-5	25
50	-1	1	46	-5	25
51	0	0	47	-4	16
47	-4	$\underline{16}$	43	-8	$\underline{64}$
		$\Sigma = 34$			$\Sigma = 194$

Total $SS = \Sigma d_g^2 = 576$

CARTOON MAIN EFFECT SS
—where d_c equals deviation of scores about the grand mean.

	Cartoon	
	A	N
Mean	54	48
Deviation (d_c)	$+3$	-3
d_c^2	9	9
nd_c^2	90	90

$SS = \Sigma nd_c^2 = 180$

SEX MAIN EFFECT SS
—where d_s equals deviation of sex means from the grand mean.

	Sex	
	m	f
Mean	55	47
Deviation (d_s)	$+4$	-4
d_s^2	16	16
nd_s^2	160	160

$SS = \Sigma nd_s^2 = 320$

Table 9.3 (cont'd.)

ERROR (WITHIN) SS
—where d_w equals deviation of scores from group means.

Group Am	d_w	d_w^2	Group Nm	d_w	d_w^2
	-2	4		-2	4
	-1	1		$+1$	1
	-1	1		$+1$	1
	$+2$	4		$+2$	4
	$+2$	4		-2	4
Group Af			Group Nf		
	-2	4		-2	4
	$+1$	1		$+1$	1
	$+1$	1		$+1$	1
	$+2$	4		$+2$	4
	-2	4		-2	4

$SS = \Sigma d_w^2 = 56$

INTERACTION SS
—where d_i equals deviations from the grand mean of means with main effects removed.

	Group			
	Am	Af	Nm	Nf
Original mean:	59	49	51	45
Cartoon effect removed:	56	46	54	48
Sex effect removed:	52	50	50	52
Deviation (d_i) from grand mean:	$+1$	-1	-1	$+1$
d_i^e:	1	1	1	1
nd_i^2:	5	5	5	5

$SS = \Sigma nd_i^2 = 20$

SUMMARY TABLE

Source	SS	d.f.	MS	F*
Cartoons	180	1	180.0	51.43
Sex	320	1	320.0	91.43
Cartoon × sex	20	1	20.0	5.71
Error	56	16	3.5	
Total	576	19		

* MS of effect ÷ MS error.

is due to sampling error. For the present data, these F ratios and their associated probability values are as follows:

Cartoon main effect (testing N.H.: $\mu_N = \mu_A$); $F = 51.43$; with $d.f. = 1/16$, $p < .01$.

Sex main effect (testing N.H.: $\mu_m = \mu_f$); $F = 91.43$; with $d.f. = 1/16$, $p < .01$.

Interaction effects (testing N.H.: $[\mu_{Nm} - \mu_{Am}] = [\mu_{Nf} - \mu_{Af}]$; $F = 5.71$; with $d.f. = 1/16$, $p < .05$.

Results

In the present statistical results, there are grounds for rejecting all three null hypotheses because in each case the calculated value of probability reached the level required for significance. Because the interaction hypothesis was of main concern in the study, the conclusion would center upon this outcome:

> There is evidence that the tendencies to engage in aggressive play after viewing a televised cartoon portraying aggression as opposed to a cartoon not portraying aggression are different for male and female children of kindergarten age.

Obviously, there is more to say about the results than is stated above. But such further interpretations will in most cases depend on subsequent comparisons of individual means (discussed later in this chapter).

F RATIOS IN A MULTIPLE-FACTOR MODEL

Most of the logic in calculating F ratios for the present multiple-factor model is not much different from what was discussed for the single-factor model. Again, the numerator of this ratio represents in each case a kind of between-groups comparison, whereas the denominator is a variance estimate of the single population expected under the terms of the null hypothesis. Let us consider the nature of these sources of variation in the present example.

Total Variation

Just as in the calculation of F in Chapter 8, we can consider the concept of total variation of all scores about the grand mean. Examine this calculation in Table 9.2. The grand mean is 51. The total sum of squares, or the sum of all squared deviations of scores about the grand mean, is 576, as shown in Table 9.3. This total sum of squares has "everything in the pot." Some of the variation represents main effects of the cartoon factor; some of it comes from the sex factor. Still more variation may come from what special effects the two factors have when combined, that is, interaction effect. Finally, some of the variation represents individual differences among the subjects; this is referred to as *error* variance. In the use of a multiple-factor analysis of variance, we shall be partitioning this total variation according to the above sources. In the particular analysis model to be used, the variation expected under the null hypothesis, and the denominator of the F ratios, is again the concept of variance *within* groups, or *error* variance. Each of the other sources of variation, by contrast, will contribute to the numerator of a series of F ratios—one for each main effect, one for the interaction.

Main Effects

Note in Table 9.3 how the sources of variation due to the two main effects (cartoons, sex) are calculated. Similar to the between-groups source of variation in the single-factor model, each of these main effects reflects deviation of the two means of the levels of each factor from the grand mean. Each main effect is treated as if the other main effect did not exist. That is, for example, when the cartoon main effect is calculated, all 10 scores for each of the two cartoon groups (the two levels) are taken into account. There is no distinction made for the fact that half of these scores came from boys and the other half from girls. In short, the cartoon main effect is calculated as if there were just two groups in the experiment, each one corresponding to a level of the cartoon factor. The comparison of cartoon means ($M_A = 54$, $M_N = 48$) is then conducted by use of an F ratio that has a main-effects variance in the numerator (180) and the error variance (3.5) in the denominator. The main effect of sex is treated in the same manner. The two levels of this factor ($M_m = 55$, $M_f = 47$) have a main-effects source of variation (320), and it too is cast into an F ratio with the error variance (3.5).

Interaction Effects

Strictly speaking, variation due to interaction effects is only that variation that is *not* attributable to either of the main effects nor treated as error variance. That is, interaction variance is whatever nonerror variation is observed among the individual group (Nf, Nm, Af, Am) when the main-effects variation has been removed. Put another way, interaction effects are not the sum of the main-effects combination but are the nonerror effects that occurred above and beyond this sum. In other words, inter-action centers upon the question: What special additional effects are due to the combination of factors? This definition of interaction effects can best be seen by examining the calculation shown in Table 9.3.

Note the segment of Table 9.3 devoted to calculation of interaction variance. Although this is not the typical way such variance is calculated (there are short-cut methods), it does illustrate just what is being consid-ered in terms of interaction variance.

As was said above, the interaction variance is whatever remains in the way of group differences after the main-effects variations have been re-moved. Suppose, then, that we started with the means in terms of the actual outcome of the study; examine these in Table 9.2 or 9.3. What are the consequences on the four individual group means if we remove the main effect of cartoon difference? Examine Table 9.4. In this case, the two marginal means for cartoon effects have been equalized and, in turn, are equal to the grand mean. Making such an adjustment entailed reducing each of the individual group means for cartoon A by 3 points and increasing each of the group means for cartoon N by 3 points. The effect was to remove all difference among the means that was due to the cartoon factor. The consequences to the individual group means are shown in Table 9.4. Next, we remove the main effects of the sex factor. This entails adjusting group means until the two sex means are equal to each other as well as equal to the grand mean. The consequences of this upon individual group

Table 9.4   Consequences of Removing Main-effects Sources of Variation

Cartoon Effect Removed:

		Cartoon		
		N	A	
	m	54	56	(55)
Sex				
	f	48	46	(47)
		(51)	(51)	

Cartoon means are now equal

Sex Effect Removed:

		Cartoon		
		N	A	
	m	50	52	(51)
Sex				
	f	52	50	(51)
		(51)	(51)	

Sex means are now also equal

means are again shown in Table 9.4. What remains are the group means (50, 52, 52, 50), which serve in the calculation of interaction variance.

If there were no interaction effects whatsoever (and no sampling error), we would expect each of these cells to equal the grand mean (that is, 51). But there are some differences remaining, and the question is whether their variation about the grand mean is significant. We calculate the variation (see Table 9.3) of these interaction means about the grand mean in the same way as the main-effects calculation. In the present example, the sum of squares is 20, as is the variance. This, too, is placed in an F ratio with the error variance and subsequently interpreted for a value of probability of occurrence under the null hypothesis.

Interpretations

The different components of variation are usually placed in a summary table similar to the one shown in Table 9.3. This table also reports the F ratios of interest and their associated probabilities. Again, the probability of each value of F occurring under the terms of the null hypothesis is obtained by consulting a sampling distribution of F (for instance, Table 8.2).

Interpretation of main effects is relatively straightforward. In the present sample study, each main-effect comparison (for example, M_A versus M_N, M_m versus M_f) can be assessed from the outcome of the probability associated with the value of F. Because the probability was sufficiently low for rejection of the two null hypotheses, this in itself was enough to provide a basis for acceptance of the two main-effects research hypotheses and practical interpretation of the results. Had it been the case, however, that

Table 9.5   **Tabular Report of the Interaction Interpretation in the Sample Study**

		Cartoon	
		N	A
	m	51_b	59_c
Sex			
	f	45_a	49_b

(Means with common subscripts
are not significantly [$p < .05$]
different from one another.)

one or both factors had more than two levels, then it might have been necessary to make some subsequent statistical comparisons between in- dividual means. Only by such comparisons would it be possible to find precisely what mean differences were contributing to an overall difference. These comparisons would entail much the same procedure as was de- scribed in Chapter 8 where there were three cartoons involved in the experiment.

The interpretation of an interaction does not necessarily stop with the appropriate F test. Suppose that, in the present case, the interaction had not been significant. Such an outcome would be clear-cut; we could not reject the null hypothesis posed for the interaction and, hence, would have no support for the hypothesized interaction. By contrast, however, because the interaction is significant, what kind of support does this give us? The significant interaction tells us that whatever differences are ob- served between cartoon levels A and N under one level of sex (for example, male) are not the same as found under the second level of sex (female). This in itself is a statement of the logical alternative to the null hypothesis, that is, the interaction research hypothesis that has been posed.

In order to determine, however, whether the outcome of the study conforms to our expected differences in terms of the relative effects of the cartoons upon boys and girls, it is necessary to examine further dif- ferences among individual group means. Such differences are tested by the use of the same methods of multiple mean comparisons as were dis- cussed in Chapter 8. One test of this type[1] would lead to the outcome reported in Table 9.5 (calculations not shown). This puts the researcher in a position to state the results in considerably greater detail; for example:

There is evidence that male children of kindergarten age have a greater tendency to engage in aggressive play than do female children of the same age when both have been exposed to a televised cartoon depicting aggression compared with a cartoon showing no aggression.

[1] D. B. Duncan, "Multiple Range and Multiple F Tests," *Biometrics*, XI (1955), 1–42.

Interaction Patterns

Earlier, it was said that for the sample study a nonsignificant interaction would indicate that the difference in effects of the two cartoons (A, N) on males would be the same as the difference in effects of the two cartoons on females. By the same token, a nonsignificant interaction also indicates

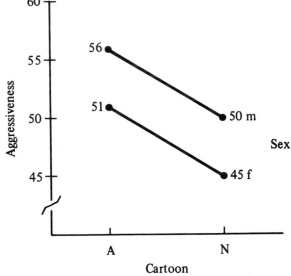

FIGURE 9.1   A nonsignificant interaction

that the difference between the sexes (m, f) in effects of one cartoon (such as A) would be the same as the difference between the sexes in effects of the alternative cartoon (N). In reality, the foregoing two statements are simply two ways of saying the same thing in terms of the conditions that would underlie a nonsignificant interaction. To portray this point, consider Figure 9.1, two graphs of the same nonsignificant interaction. These two graphs illustrate the characteristic feature of a nonsignificant interaction: The lines are parallel, or near parallel.

By contrast, consider Figure 9.2, a graph of the significant interaction found in the sample study. Note how the lines are not parallel. The difference between cartoons A and N is greater for males (m) than for females (f). The fact, however, that an interaction is significant is not contingent on this particular pattern of differences among the individual group means. Consider, for example, Figures 9.3 and 9.4. Both illustrate at least two patterns that would also underlie a significant interaction. In Figure 9.3, the pattern indicates an exact reversal of differences between the two cartoons for each sex. On cartoon A, the male group is lower in aggressive tendencies than on cartoon N, but just the opposite prevails for the female group. In Figure 9.4, the two cartoons have no differential effect on females, but the males show more aggressive tendencies after viewing cartoon A than after viewing cartoon N.

More patterns than shown here could illustrate cases of significant interactions. The point in showing at least some of the different patterns is that a significant interaction, in itself, is usually not sufficient as a basis

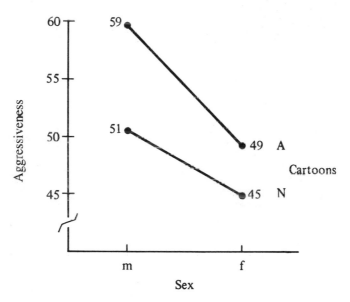

FIGURE 9.2   The interaction found in the sample study

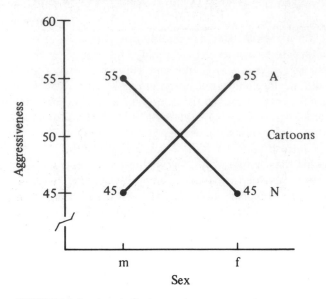

FIGURE 9.3   A significant interaction

for interpreting the detailed results. What is necessary is an interpretation—often with subsequent comparisons of the individual means—of the interaction pattern.

A final point concerns what consequences a significant interaction has on interpretation of main effects. As we have been discussing all along,

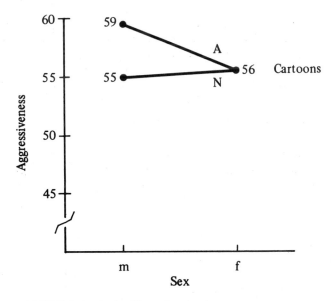

FIGURE 9.4   A significant interaction

a significant interaction indicates that whatever differences are observed between (or among) levels of one factor are not consistent across levels of another factor. Obviously, this often places conditions on how the main effects are interpreted. In the sample study, for example, although male children consistently had a greater tendency to engage in aggressive play than did females, this difference was greater after viewing cartoon A than after viewing cartoon N. Or consider the interaction portrayed in Figure 9.3. Here the main effects of the cartoon factor, as well as the sex factor, would be nonsignificant (all means of the cartoon and sex main effects would be equal to 50). Interpretation in this case of only the main effects would be quite misleading. The experiment does indeed have significant effects, but these are solely in the interaction.

The rule of thumb is usually this: If there is a significant interaction, then the interpretation of each of the main effects must take into account the consequences of the interaction. On the other hand, if the interaction is not significant, then the main effects can be treated as independent of one another—that is, they can be interpreted individually, without being somehow conditional on one another.

FURTHER NOTES ON MULTIPLE-FACTOR MODELS

As in the discussion of single-factor models, we have only touched the surface in describing a multiple-factor model. We can, however, introduce some additional points that are characteristic of more complex designs.

Levels and Factors

Like a single-factor model, the levels of individual factors in a multiple-factor design represent a compromise between theory and design. The sample experiment incorporated a 2 × 2 model—that is, two factors, each with two levels. Suppose, however, that we had incorporated four cartoons into the design, with the two levels of sex. In this case, we would have a 4 × 2 analysis model, that is, four levels of the cartoon factor and two levels of the sex factor.

Also, we can increase the number of factors in a design. Suppose that the sample experiment were expanded to incorporate not only the two levels of the cartoon factor and the two levels of the sex factor but also two levels of a factor characterizing the time delay between viewing the cartoon and measurement of aggressiveness in play. This would lead to a 2 × 2 × 2 factorial model. Moreover, suppose that two levels of a factor corresponding to the children's ages were also incorporated into the design. Here we would have a 2 × 2 × 2 × 2 model, that is, cartoon × sex × time × age. Finally, if we had four, rather than two, levels of the cartoon factor and if we kept all of the above factors, the model would be described as a 4 × 2 × 2 × 2 design.

The multiple-factor models that we have looked at, thus far, show that for any given set of factors there are all possible combinations of the levels of each factor with the levels of another factor. Without going into detail, let us say that there are multiple-factor models that for one reason or another do not have this characteristic. One such design, called a *Latin square*, represents a systematic partial combination of levels of different factors. Latin squares are used mostly when a complete combination of all levels would be uneconomical. Another type of design (*nested*) represents a case where, for example, the two levels of a factor have nested within them different levels of another factor.

Higher-Order Interactions

Any time a factorial model incorporates more than two factors, it will incorporate an additional number of possible sources of interaction. Consider, for example, a $2 \times 2 \times 2$ factorial model corresponding to factors of cartoon, sex, and time delay between viewing and measurement. The model for analysis would accommodate main-effects comparisons of the two levels of the cartoon factor, the two levels of the sex factor, and the two levels of the time factor. Interactions would include:

1. cartoon $\times$ sex
2. cartoon $\times$ time
3. sex $\times$ time
4. cartoon $\times$ sex $\times$ time

The first of these interactions (1) asks whether above and beyond the main effects of the cartoon and the sex factors, there is additional variation due solely to the combination of these two factors. Again, this is not the additive effects of cartoon and sex factors but the unique effect that their combination makes. Similarly, the second and third of these interactions (2, 3) ask whether the combinations of cartoon and time factors, or sex and time factors, contribute to any unique effects. Finally, there is the three-way interaction (4), which asks whether above and beyond any main effects or two-way interaction effects, there is any special effect due solely to particular three-way combinations of the factors.

Needless to say, interactions become more and more complex, both in terms of computation and interpretation, as more and more factors are incorporated into a single design. In any two-factor design, there will only be one interaction. But, for example, in a $2 \times 2 \times 2 \times 2$ model, there are up to 10 interactions that may be necessary to analyze.

Error Terms

Another important consideration in multiple-factor models is the basis for gaining the best estimate of error variance for a particular F ratio. Recall that this estimate of error becomes the denominator in a ratio with whatever source of variation is being subjected to a hypothesis test. In relatively complex models, the selection of the appropriate estimate of error (usually called the *error term*) becomes a relatively subjective decision. Within a given model for analysis, it is often the case that a number of different error terms will be used, depending on what particular sources of variation are being tested. Such considerations as whether factors are fixed or random and whether measures are repeated or independent enter into the decision regarding the appropriate error term.

Other Variations

As is subsequently discussed in chapters on multiple regression (Chapter 13), and especially in those on multiple-variable statistics (Chapters 14 and 16), there are common mathematical properties that relate analysis of variance to other statistical models, particularly multiple regression analysis. It is possible, for example, to take parts of the sample study in the present chapter and to calculate *F* ratios by another approach, multiple regression, which centers on *relationships* between independent and dependent variables (Chapter 13). In fact, when you consider this alternative, you may see how many analyses of differences and analyses of relationships are close to reflecting the same thing: how one set of variables varies relative to another set.

Finally, as mentioned briefly in the previous chapter, analysis of variance can be done with multiple dependent variables, a technique known as multivariate analysis of variance (Kerlinger and Pedhazur, 1973).

SUMMARY

Multiple-factor analysis of variance is a statistical model for testing the consequences of manipulating two or more independent variables in a single research design. Each independent variable (factor) will have two or more levels. The *F* ratio is the statistic used to conduct the appropriate hypothesis tests in multiple-factor designs. Significance tests among different levels of each factor are known as *main effects*. Whatever effects are due solely to the combination of factors are known as *interaction effects*. Given a significant interaction, it is usually necessary not only to conduct follow-up tests of means in order to interpret results, but it is also usually necessary to impose particular conditions on the interpretations of whatever main effects have been observed.

SUPPLEMENTARY READINGS

Bochner, Arthur P., and Mary Anne Fitzpatrick, "Multivariate analysis of variance: Techniques, models, and applications in communication research," Chapter 5 in Peter Monge and J. Cappella (eds.), *Multivariate Techniques in Communication Research*. New York: Academic Press, 1978.

Howell, David C., *Fundamental Statistics for the Behavioral Sciences*. Boston: PWS-Kent, 1989. See Chapters 17 and 18.

Kerlinger, Fred N., and E. F. Pedhazur, *Multiple Regression in Behavioral Research*. New York: Holt, Rinehart and Winston, 1973.

SAS Institute, Inc., *SAS User's Guide: Statistics*. 1982 ed. Cary, N.C.: SAS Institute, 1982. See the section on "Analysis of Variance" (pp. 113–255).

SPSS, Inc., *SPSS-X User's Guide*. 3rd ed. Chicago: SPSS, 1988. Chapter 20 describes computer programs for analysis of variance.

Winer, B. J., *Statistical Principles in Experimental Design*. 2nd ed. New York: McGraw-Hill, 1971. Comprehensive approach to multiple-factor analysis of variance.

10

Nonparametric Tests

In addition to the tests discussed in the previous three chapters for assessing differences, there are tests that do not directly incorporate estimates pertaining to population characteristics. These are aptly called nonparametric tests. In this chapter, one such test, chi square, receives the most attention, although a number of other nonparametric tests of differences, including "log-linear" models, are briefly mentioned.

There are a variety of ways by which nonparametric tests can be classified. Perhaps it would be most practical to examine some of these tests in terms of the type of measurement typically involved in their usage. This latter point is one distinguishing feature of nonparametric tests. They are often used when measurement involves nominal or ordinal scaling.

WHEN DATA ARE IN CATEGORIES: USE OF CHI-SQUARE ANALYSIS

As discussed in Chapter 3, there are situations when measurement involves nothing more than assigning observations to different categories in a set of well-defined, mutually exclusive categories. Usually, in this situation, called *nominal* scaling, we are either interested in comparing categories among themselves, or we are interested in contrasting how samples differ in terms of assignment into the categories. The typical statistical model used in such cases, and one having widespread utility in communications and education research, is chi-square analysis. (Particular values of chi square are usually identified by the symbol χ^2.)

In essence, χ^2 is best thought of as a discrepancy statistic. That is, its calculation is based on the discrepancy between the frequencies observed for a set of categories and some alternative theoretical set of frequencies posed by the researcher. Like t and F, χ^2 has a sampling distribution by which we can estimate the probability that a given value of χ^2 would be expected under the terms of the null hypothesis. Chi-square analysis can be applied in a great variety of situations. We shall discuss some of the most typical of these.

Differences among Categories

Sometimes chi-square analysis is used not to compare different samples but simply to investigate how items within a sample will distribute in a set of categories. This is usually called the *one-sample* case of chi-square analysis.

Consider that a single sample of 30 teachers has been selected to evaluate three alternative versions (I, II, III) of a math lesson. Their task is to designate which one of the three lessons they consider to be the best planned. The measurement outcome of this study is designated in terms of the number of teachers judging each of the versions as best planned. These would make up the *observed frequencies*. The researcher's hypothesis is that teachers will not choose equally among the three versions; that is, one or two versions will be selected over others as being the best. Such a research hypothesis implies the null hypothesis that teachers will choose equally among the three versions. The level for rejection of the null hypothesis is set at $p < .05$. Under the null hypothesis, we would expect each version to have 10 teachers rate it as the best planned; we call these expectations the *theoretical frequencies*. Suppose that the following observations were obtained:

    Version I: chosen by four teachers as best planned
    Version II: 19
    Version III: seven

As mentioned earlier, the basis for calculating χ^2 is in the discrepancy between the observed and the theoretical distributions. The greater this discrepancy, the greater is the value of χ^2. Note in Table 10.1 how χ^2 is calculated for the present example. Here, $\chi^2 = 12.6$, and when interpreted in a chi-square sampling distribution (Table 10.2), this value is associated

Table 10.1   **Calculation of χ^2; Single-Sample Case**

		Versions		
	I	II	III	(total)
Observed freq.	4	19	7	30
Theoretical freq.	10	10	10	30

$$\chi^2 = \Sigma \left[\frac{(O - T)^2}{T} \right] = \frac{(4 - 10)^2}{10} + \frac{(19 - 10)^2}{10} + \frac{(7 - 10)^2}{10} = 12.6$$

With $d.f. = 2$, χ^2 of 12.6 is associated with $p < .01$ (see Table 10.2).

Table 10.2   **Example of a Table of Chi-Square Sampling Distribution**
                        (*See a complete table, p. 209.*)

Probability	.10	.05	.02	.01
d.f.				
1	2.706	3.841	5.412	6.635
2	4.605	5.991	7.824	9.210
. . .	. . .	. . .	. . .	. . .
15	22.307	24.996	28.259	30.578
. . .	. . .	. . .	. . .	. . .
30	40.256	43.773	47.962	50.892

with $p < .01$. What this means is that the value of χ^2 would be expected in fewer than one of 100 cases of random sampling. Thus, the researcher could reject the null hypothesis (no difference among the versions) in favor of the research hypothesis (a difference in choices of the versions).

Differences Among Samples

Consider a study similar to the one just described, but this time, the research interest centers on whether males and females have different preferences in judging which of the three lessons is best planned. Rather than comparing the versions among themselves, males and females are now being compared in terms of their ratings of the three versions. Suppose that the research hypothesis holds that the two groups will have different judgments. The following data are obtained:

	Males	Females
Version I:	5	5
Version II:	5	20
Version III:	10	5

The above figures provide only the observed frequencies. Given the research hypothesis that the two samples (males, females) will be different, the implied null hypothesis is that they will not be different. The significance level of the chi-square test is set at $p < .05$. The theoretical frequencies, then, must represent this alternative, or null, hypothesis of no difference. The set of theoretical frequencies for the males and a (separate) set for the females must be calculated. This can be done by a relatively simple procedure (see Table 10.3) whereby each theoretical frequency is the product of the appropriate row sum multiplied by the column sum, divided by the grand sum. For example, the theoretical frequency for males choosing Version I is 20 (row sum) multiplied by 10 (column sum), divided by 50 (grand sum). These theoretical frequencies are entered in Table 10.3. For a given problem, each theoretical frequency accommodates differences among the categories but adjusts this to conform with the null hypothesis of no differences between the samples.

A value of χ^2 is then calculated on the basis of the discrepancies between each observed frequency and its corresponding theoretical frequency. In this case, the total $\chi^2 = 9.03$ (Table 10.3). When interpreted in a sampling distribution (Table 10.2), this value is associated with $.01 < p < .02$. Thus, the researcher rejects the null hypothesis (no difference between the samples) in favor of the research hypothesis (a difference between the samples).

Chi-square analysis can apply to more than two samples in a single test. Thus, for example, a study might involve comparison of persons of five different experience levels in judging the aforementioned versions of a math lesson. Chi-square analysis can also involve more categories than shown in the foregoing examples. Perhaps there are five or even 10 versions

Table 10.3   Calculation of χ^2; Multiple-Sample Case

| | Versions | | | |
	I	II	III	Row Sums
Males:				
Observed freq.	5	5	10	20
Theoretical freq.	4	10	6	(20)
Females:				
Observed freq.	5	20	5	30
Theoretical freq.	6	15	9	(30)
Column sums	(10)	(25)	(15)	Grand sum = ((50))

$$\chi^2 = \Sigma \left[\frac{(O - T)^2}{T} \right] \frac{(5 - 4)^2}{4} + \frac{(5 - 10)^2}{10} + \frac{(10 - 6)^2}{6}$$

$$+ \frac{(5 - 6)^2}{6} + \frac{(20 - 15)^2}{15} + \frac{(5 - 9)^2}{9} = 9.03$$

With $d.f. = 2$, χ^2 of 9.01 is associated with $.01 < p < .02$ (see Table 10.2).

of the math lesson; these could all be incorporated into a chi-square analysis. Or, for that matter, chi-square analysis could apply to a design where five different experience level groups judged 10 versions of the lessons.

In the sense just discussed, chi-square analysis is often seen as a test of the goodness of fit between two distributions (considering distributions in terms of nominal measurement categories). Suppose, for example, that we have drawn samples of students from two universities and we are concerned with whether the two samples have students distributed equally in the same age categories. In other words, the question concerns the goodness of fit between the two distributions in terms of frequencies in various age groups. A nonsignificant χ^2 value would support the assumption that the two distributions fit one another, whereas a significant χ^2 value would indicate lack of fit.

Calculation of χ^2

The general formula for the calculation of values of χ^2 is given in both Tables 10.1 and 10.3. Just how the formula is applied is also seen in these two tables. For example, in the single-sample use of chi-square analysis, there were three cases of $O - T$. Each of these enters into the calculation of what appears between the brackets in the formula; then, the results of the three cases are summed to yield a total value of χ^2. Similarly, in the example where two samples were compared in terms of selection from among the three versions of the lesson, all $O - T$ cases are considered; only this time, there are six cases to sum across in order to yield χ^2.

Interpretation of χ^2

Values of χ^2 are interpreted for values of associated probability in terms of a sampling distribution posed for the null hypothesis. In practice, this sampling distribution is given in tabular form similar to the example in Table 10.2. Here, it is necessary to enter the table with a value known for degrees of freedom ($d.f.$). In the case of chi-square analysis for a single sample, degrees of freedom is equal to the number of categories considered minus one. Thus, in the earlier example (Table 10.1), degrees of freedom is equal to $3 - 1$, or 2. When samples are being compared, degrees of freedom is equal to the number of samples minus one being compared multiplied by the number of categories minus one. In the foregoing example (Table 10.3), this was equal to $(2 - 1) \times (3 - 1)$, or 2.

As can be seen in Table 10.2, for given degrees of freedom, the larger the value of χ^2, the lower is the associated value of probability. This is a reflection of the fact that as the discrepancies between observed (O) and theoretical (T) frequencies increase, so does the value of χ^2, and the less is the probability of occurrence under a hypothesis of no difference.

Log-Linear Models

Although chi-square calculations are limited to one- and two-way models, it is possible to analyze categorical data incorporating further dimensions by a technique known as *log-linear* models. Suppose, for example, that categories of readers and nonreaders of a newspaper were further subdivided by three categories of age (under 30, 30–60, over 60), and two of sex. Using two-way chi-square analyses (that is, rows by columns), it would be possible to assess the relationship (or independence) of any two of these variables but not all three simultaneously. Using a log-linear model, which also reflects patterns of observed and expected frequencies, it is possible to test the significances of differences associated with each of the three primary variables as well as their four interactions (for example, readership by age and sex by age). Thus, with only categorical data, it is possible to test higher-order interactions similar to those usually found in complex analysis of variance or regression models.

The *SPSS-X User's Guide* (see Supplementary Readings to this chapter) will provide you with a practical example of calculation of log-linear models.

WHEN DATA ARE RANK ORDERED: OTHER NONPARAMETRIC TESTS

In Chapter 3, the concept of *ordinal* scaling was introduced. In this scale, we know nothing about the measurement intervals, but we do know that the categories have an ordered relationship. That is, we do know that one category represents a greater degree of some characteristic than another category. A very practical shortcoming when using an ordinal scale is that without a knowledge of measurement intervals there are limitations on the arithmetic operations that we can perform using the numbers assigned to these scales. This obviously imposes restrictions on the types of statistical models that can be used when ordinal scaling is involved, that is, unless the researcher is willing to hazard whatever error is involved in assuming that a scale has interval properties.

The statistical models that can be applied to ordinal scaling, in one way or another, stay within the limitations imposed by this level of measurement. In briefly reviewing some of these models, notice how they accommodate such limitations. Detailed descriptions of the calculations involved in the models to be discussed are found in Siegel's *Nonparametric Statistics for the Behavioral Sciences*.

Independent Measures

We shall consider first the case where we have two groups of measures that are unrelated (that is, they are measures of different Ss). For purposes of comparison, we shall use the data from the sample study discussed in Chapter 7 where one group (I) of children was exposed to a televised cartoon depicting aggression, whereas a second group (II) was exposed to a cartoon showing no aggression. The dependent variable comprises judgmental ratings of the children's tendency to engage in aggressive play after seeing one or the other of the cartoons. In the present case, however, we shall consider that the judges' ratings have only the power of an *ordinal* scale. It will now be necessary to assume that the average of judges' ratings for each child is now a *median* rather than a mean (because we cannot perform the arithmetic calculations necessary for a mean). These median ratings are as follows:

Group I	Group II
61	50
53	48
59	56
55	54
57	52

The hypothesis to be tested can no longer be stated in terms of mean scores; it will have to be adapted to the nature of the statistical test to be used. However, the research hypothesis is still that the two groups will be differentially affected by the two cartoons. Some of the nonparametric tests that might be applied are next briefly discussed.

The *median test* can be used to test the research hypothesis that the two groups come from populations having different medians. The test itself, however, centers upon the null hypothesis that the groups come from populations having the same median. To understand how a probability value is calculated for this null hypothesis, examine how the data are displayed in Table 10.4. A median for all 10 of the scores (both groups

Table 10.4   **Example of the Median Test**

Grand Median of Combined Distributions = 54.5

	Cartoons	
	I	II
	(A)	(B)
Ratings above grand median	4	1
	(C)	(D)
Ratings below grand median	1	4

Probability Calculation*

$$p = \frac{(A + B)!(C + D)!(A + C)!(B + D)!}{N!A!B!C!D!}$$

$$p = \frac{(5)!(5)!(5)!(5)!}{(10)!(4)!(1)!(1)!(4)!}$$

$$p = \frac{(120)(120)(120)(120)}{(3,628,800)(24)(1)(1)(24)}$$

$$p = .099$$

---

* "!" is the notation for a factorial. For example, 5! is read as "factorial 5" and means $1 \times 2 \times 3 \times 4$ $\times$ 5 = 120.

combined) is found· In this case, it is 54.5. The group scores are then distributed in a 2 × 2 table that indicates the number of scores in each group that are above and below this overall median. If the two groups had exactly the same median and if there were no sampling error whatsoever, we would expect this division of scores to be the same for both groups. This would be the expectation under the terms of the null hypothesis. When a probability (p) value is calculated by use of an appropriate formula (Table 10.4), it indicates the probability that whatever discrepancies have been observed between the two groups could occur under the terms of the null hypothesis.

As applied to the same data, the *Mann-Whitney U test* does not test precisely the same hypothesis. The research hypothesis in this case says that the two groups come from populations having different distributions— or in terms of the null hypothesis, that the two groups have the same distribution. Table 10.5 illustrates how this test would be applied. The scores for both groups are ranked together from lowest to highest while retaining a label to indicate what group each score came from. The value of U reflects the number of times scores for one group precede scores for the other group in this distribution. In the example in Table 10.5, no score for Group II is preceded by a Group I score until the score 54; this contributes a value of 1 to the calculation of U. At the score 56, a Group II score is preceded now by two Group I scores. After the whole distri-

Table 10.5   Example of the Mann-Whitney U Test

Score Ranking:

Score:	48	50	52	53	54	55	56	57	59	61
Cartoon:	(II)	(II)	(II)	(I)	(II)	(I)	(II)	(I)	(I)	(I)
U:	0	0	0	0	1	0	2	0	0	0 = 3

U = the total number of times scores for Group I
precede scores for Group II

Probability of obtaining the above is $<.056$ (by use of tabled values).

Table 10.6   Example of the Kolmogorov-Smirnov Test

Cumulative Frequency Distribution:

Score Categories	Cartoons I	II	
61	5		
60	4		
59	4		
58	3		
57	3		
56	2	5	→ discrepancy = 3
55	2	4	
54	1	4	→ discrepancy = 3
53	1	3	
52		3	
51		2	
50		2	
49		1	
48		1	

Probability of obtaining the above is $>.05$ (by use of tabled values).

bution is examined, the total value of U is found to be 3. A probability value is then determined (usually by appropriate tables) that indicates the probability that a value of U of this magnitude or smaller could occur under the terms of the null hypothesis.

The *Kolmogorov-Smirnov test* also tests the research hypothesis that two distributions differ. Table 10.6 illustrates how this test would be applied to the present data. Cumulative frequency distributions[1] of the scores in each group are prepared. The interval of greatest discrepancy

---

[1] In working up from the lowest score interval (Table 10.6), a cumulative frequency distribution shows the sum of the score frequencies up to and including each particular interval.

between these two distributions then yields a particular value (two intervals in this case, both of them with a discrepancy of 3). A probability value is then determined (by calculation or from an appropriate table) that indicates the probability of such a discrepancy occurring under terms of the null hypothesis.

Repeated Measures

When two sets of scores are somehow related (that is, measures on the same Ss, or matched Ss), different nonparametric models are used. To illustrate two of these models briefly, consider that we have two sets of measures taken on the same six Ss. One set comprises measures (ordinal scale) of attitude toward blood donation taken prior to exposing the Ss to a 5-minute film encouraging such donation. The second set comprises postexposure measures. The research hypothesis, in a general form, is that the pre- and postexposure measures represent different conditions in terms of attitude toward blood donation (or more practically speaking, that the film has created a different condition of attitude). The ratings are as follows:

(S)	*Preexposure*	*Postexposure*
1	8	10
2	6	8
3	7	9
4	5	8
5	4	7
6	4	5

Table 10.7 illustrates how the *sign test* could be applied to the above data. Each rating pair is assigned either a plus ($+$) or a minus ($-$), indicating in this case whether the postexposure score exceeds the preexposure score

Table 10.7   **Example of the Sign Test**

Directional Sign of Preexposure versus Postexposure Ratings:

	Ratings		
S	Preexposure	Postexposure	Sign
1	8	10	+
2	6	8	+
3	7	9	+
4	5	8	+
5	4	7	+
6	4	5	+

Probability of obtaining the above is $< .05$ (by use of tabled values).

Table 10.8   **Example of the Wilcoxon Matched-Pairs Signed-Ranks Test**

Magnitude and Sign of Score Differences:

| | Ratings | | | Rank of Differences |
| | Preexposure | Postexposure | Differences | + — |
S				
1	8	10	+2	3.0
2	6	8	+2	3.0
3	7	9	+2	3.0
4	5	8	+3	5.5
5	4	7	+3	5.5
6	4	5	+1	1.0

Σ of less frequent
rank $(-) = 0$

Probability of obtaining the above is $<.05$ (by use of tabled values).

(thus, is marked $+$). The null hypothesis poses that the two sets of ratings are the same. If this were the case, we would expect roughly an equal number of pluses and minuses to occur. Or in other words, as the two conditions differ, we would expect to find a preponderance of one or the other sign. The value of probability in this case refers to the probability that we would expect a particular discrepancy between plus and minus signs to occur under the terms of the null hypothesis.

As shown in Table 10.8, notice that the *Wilcoxon matched-pairs signed-ranks test* takes into account the magnitude of the difference between rankings of the scores in the two distributions. Similar to the sign test, these differences (now actual values of rank differences) are given plus and minus signs. If the null hypothesis of no difference between the two conditions were the case, we would expect the sum of the plus differences to equal roughly the sum of the minus differences. Again, a value of probability indicates the probability under the terms of the null hypothesis of obtaining a particular discrepancy between the sums of the plus and minus differences.

FURTHER NOTES ON NONPARAMETRIC TESTS

Other Models

What we have covered thus far are only some of the more frequently used nonparametric tests. There are a number of additional tests of differences that may be more appropriate for certain research designs than the tests discussed in this chapter. For example, there are nonparametric tests that may apply when the researcher wishes to assess differences among more than two groups. Typically, the *Friedman two-way analysis of variance* is used when repeated measures are involved, and the *Kruskal-Wallis one-*

way analysis of variance is used for independent samples. Finally, there are nonparametric models used for tests of relationship. One of these, *Spearman's rho,* is discussed in Chapter 11.

Power

The concept of statistical power was previously introduced in Chapter 6. Again, what it reflects is the probability of rejecting a null hypothesis that is, in fact, false. The concept of power is particularly appropriate when considering nonparametric models because they are considered as generally less powerful than their parametric counterparts. What this means in practical terms is that, for a given set of data, the probability of rejecting a false null hypothesis is usually slightly better when using a test such as *t,* as compared with, say, the *median test,* the *Mann-Whitney U test,* or the *Kolmogorov-Smirnov test.* Or in terms of error, there is usually a slightly greater chance of committing Type II error (failure to reject a false null hypothesis) when using a nonparametric test, all other factors being equal.

Some illustration of the above point can be seen by comparing the obtained probability values reported in the present chapter with the probability value obtained with the *t* test on the same data in Chapter 7. The difference was that the probability values obtained from the use of the three nonparametric models were all slightly greater than that obtained by use of the *t* test. Thus, for example, if the level set for rejection of the null hypothesis is $p < .05$, the various tests applied to the same data would lead to different conclusions in some cases.

Although the argument over the slightly lower power of nonparametric tests is a sound conceptual point, it is not always a practical one. First, much depends on the kinds of error that the researcher is willing to tolerate. Thus, for example, it may be more to a researcher's advantage to tolerate what are perhaps minor doubts about meeting necessary assumptions of a given test of a parametric type rather than to hazard the error possible with a nonparametric test. On the other hand, to a certain degree, the power of a given nonparametric test can be increased by increasing the size of the samples used in the study. Sometimes, too, the nature of the measurement and its distributions simply dictates the use of a nonparametric test, and there are no other alternatives. However, the decision to use a nonparametric statistical test, and even the particular type of nonparametric test, is not always an easy one. This decision, like many others in reasoning with statistics, often depends on weighing a considerable number of relevant factors.

SUMMARY

Nonparametric statistical models are generally used when the researcher is faced with the use of nominal or ordinal scaling or when particular assumptions cannot be made about the nature of the populations being studied. With nominal scaling, a commonly used statistic is *chi-square analysis*. It is applied in tests centering upon how observations in a sample distribute into different categories as well as how two samples may differ in terms of their distributions. Chi-square analysis is also often used in the sense of a test of goodness of fit between distributions. When ordinal scaling is involved, one distinction among different nonparametric tests is whether the two sets of measures are related. There are additional nonparametric tests for comparing more than two groups as well as tests of relationships among sets of measures. Generally speaking, nonparametric tests are considered less powerful than their parametric counterparts.

SUPPLEMENTARY READINGS

Haberman, Shelby J., *Analysis of Qualitative Data*. 2 vols. New York: Academic Press, 1978. Includes materials on log-linear models.

Hays, William L., *Statistics*. 3rd ed. New York: Holt, Rinehart and Winston, 1981. See Chapter 16 for various nonparametric tests.

Howell, David C., *Fundamental Statistics for the Behavioral Sciences*. Boston: PWS-Kent, 1989. See Chapters 19 and 20.

SAS Institute, Inc., *SAS User's Guide: Statistics*. 1982 ed. Cary, N.C.: SAS Institute, 1982. See the general section on chi-square (pp. 245–255) or log-linear models (pp. 257–285).

Siegel, Sidney, *Nonparametric Statistics for the Behavioral Sciences*. New York: McGraw-Hill, 1956. The classic of the field.

Spence, Janet T., John W. Cotton, Benton J. Underwood, and Carl P. Duncan, *Elementary Statistics*. 4th ed. Englewood Cliffs, N.J.: Prentice Hall, 1983. See Chapter 15.

SPSS, Inc., *SPSS-X User's Guide*. 3rd ed. Chicago: SPSS, 1988. Chapter 38 describes a program for nonparametric tests; Chapter 32 describes log-linear analysis.

PART FIVE

ANALYSES OF RELATIONSHIPS

11

Correlation

In chapters 7 through 10, we were concerned primarily with assessing differences between populations based on what we knew about differences in sample distributions. In all such cases, the question of difference was centered in terms of measurement of some particular variable. In the present chapter, and the two subsequent chapters, we turn to a different type of question to be answered statistically: To what degree do two or more variables show interrelationships in a given population? One way to assess such relationships is by statistical procedures of correlation.

THE NATURE OF CORRELATION

What is meant by correlation? Suppose that we have two sets of measurements taken on the same people. One measure consists of scores on an intelligence test; the other measure is an assessment of reading ability. Presumably, intelligence and reading ability are two traits that would tend to vary together in a given population. That is, if one person has a high score on the intelligence test, it is likely that he or she would also have a relatively high score on the reading test. Similarly, a person with a low score in intelligence would be expected also to have a low score in reading. In this case, we would say that the two tests show *positive correlation;* that is, the measures vary together.

By contrast, suppose that we again had two sets of measurements taken on the same people. One measure is again a score on a test of reading ability, but the second measure is the time that it takes a person to read a passage of a given length. Presumably, a person with a relatively high score in reading ability would spend less time in reading a passage than a person with a low score in reading ability. In this case, the two measures would be said to vary inversely; that is, as one measure is high, the other is low. This would be a case of *negative correlation.*

Correlation, then, characterizes the existence of a relationship between variables. Although there may be many reasons for a relationship, correlation says nothing about these reasons. It indicates only that two or more variables vary together either positively or negatively. This may be because one variable is the cause and the other the effect, or perhaps they both vary together as a result of a third variable being the cause. Correlation itself only indexes the degree of relationship; this index is called a *correlation coefficient.*

Coefficient of Correlation

A correlation coefficient indexes two properties of a relationship. The first of these is the *magnitude* of the relationship, that is, the degree to which the variables vary together. The second is the *direction* of the relationship, whether the variables vary together (positively) or whether they vary inversely (negative). The coefficient itself has the following range of values:

+ 1.0 (perfect positive correlation)
0.0 (no correlation)
− 1.0 (perfect negative correlation)

Thus, for example, if a correlation of + .50 is found, this generally means that there is some degree of positive relationship, but it is not a perfect one.

Significance of Correlation

In addition to interpreting the magnitude or direction of a correlation coefficient, there are methods for testing the significance of a given value of correlation. Here the research hypothesis is typically that two variables are related; that is,

(R.H.) correlation $\neq$ 0

Accordingly, the null hypothesis says that no correlation exists and that whatever value of correlation is found is due to sampling error:

(N.H.) correlation = 0

A SAMPLE STUDY

Again, the present data are simulated. Ordinarily, a correlation study would not be undertaken with a sample this small.

Problem

Research literature suggests that people who read a lot also tend to be good readers compared with people who spend very little time reading. The researcher wishes to test this statement and at the same time gain some index of the magnitude of the relationship that might be involved. There is no reasoning here concerning cause and effect, only an interest in the degree to which measures of reading ability and time devoted to reading vary together in a given population.

Two measures are defined for the study. One, defined as variable X, is a score on a reading aptitude test. The second, defined as Y, is the average number of hours that a person has spent in reading books during each week of a year. Using the symbol r to represent a coefficient of correlation, the research hypothesis is stated as:

$$r_{XY} \neq 0$$

That is, it is predicted that the variables X and Y have a correlation that differs from zero, in either a negative or a positive direction. The null hypothesis is:

$$r_{XY} = 0$$

The significance level is set at $p < .05$.

Method

After obtaining a sample of 10 adults, the researcher has them estimate the average number of hours they spend reading books each week (Variable Y), then administers to them a standard test of reading aptitude (Variable X). The data are reported in Table 11.1.

Table 11.1   **Example of the Pearson Product-Moment Correlation and a Test of Significance: Two-Variable Problem (Simulated Data)**

Calculation of r

Ss	Raw Data X (Reading Aptitude)	Y (Hours)	Deviations from Means x	y	xy	x²	y²
1	20	5	0	0	0	0	0
2	5	1	− 15	− 4	60	225	16
3	5	2	− 15	− 3	45	225	9
4	40	7	20	2	40	400	4
5	30	8	10	3	30	100	9
6	35	9	15	4	60	225	16
7	5	3	− 15	− 2	30	225	4
8	5	2	− 15	− 3	45	225	9
9	15	5	− 5	0	0	25	0
10	40	8	20	3	60	400	9
Σ	200	50			$\Sigma xy = 370$	$\Sigma x^2 = 2050$	$\Sigma y^2 = 76$
Mean	20	5					

$$r_{xy} = \frac{\Sigma xy}{\sqrt{\Sigma x^2 \cdot \Sigma y^2}} = \frac{370}{\sqrt{2050 \cdot 76}} = \frac{370}{\sqrt{155,800}} = \frac{370}{394.7} = +.937$$

Significance of r (N.H.: $r = 0$)

$$t = \frac{r\sqrt{n-2}}{\sqrt{1-r^2}} = \frac{.937\sqrt{8}}{\sqrt{1-.878}} = \frac{2.65}{\sqrt{.122}} = 7.59$$

t is interpreted in Table 7.2, using $d.f. = n - 2$. Given this interpretation, the r of $+.937$ would be expected to occur in fewer than one of 100 cases if due to sampling error (that is, $p < .01$); this, the null hypothesis of $r = 0$ is rejected.

Using a statistical method called Pearson's product-moment correlation, a value of the correlation coefficient r is calculated (solution shown in Table 11.1). In this case, $r_{XY} = .937$. For the present sample size, the researcher estimates (Table 11.1) the probability that an r of .937 could occur due to sampling error, that is, under the terms of the null hypothesis. This probability value is $p < .01$. That is, we would expect a coefficient of this magnitude to occur in fewer than one chance out of 100 on the basis of random sampling.

Results

Because the probability of obtaining $r = .937$ is less than the level set for rejection ($p < .05$), the null hypothesis is rejected in favor of the research hypothesis. The researcher concludes that the two measures—reading aptitude and time spent in reading per week—are indeed related; for example:

> The number of hours that a person spends in reading books each week is likely to have a very high positive correlation ($r = .937$ in the present study) with a measure of reading aptitude.

THE LOGIC UNDERLYING PRODUCT-MOMENT CORRELATION

Like many of the statistical concepts covered in previous chapters, correlation reflects reasoning that incorporates the concept of how scores vary within a distribution. Just how this reasoning works can be seen by considering what goes into the formula for product-moment correlation.

The Formula for r

Consider first the formula that was used for the calculations presented in Table 11.1. The x and y symbols represent deviations of the X and Y scores about the means of their two distributions. Recall (Chapter 4) that such deviations are a characteristic of the dispersion of scores in a distribution. The more dispersed are the scores about the mean, the larger these deviations will be. We shall see that a particular ratio among these deviations in the two distributions is what underlies a correlation coefficient.

Consider next what the denominator of the correlation formula represents. The product of the summed squared deviations of the two distributions (that is, $\Sigma x^2 \cdot \Sigma y^2$) is the total amount of the squared deviations that the two distributions could have maximally in common. We can represent this value as the area in a square, similar to the concept of sum of squares. We have done this in part A of Figure 11.1 in terms of the data from the same study. The length of one side of this square is the value corresponding to the square root of this area; this root value, in turn, corresponds to what we calculated for the denominator of the correlation formula.

$\sqrt{\Sigma x^2 \cdot \Sigma y^2} =$
394.7

$\Sigma x^2 \cdot \Sigma y^2 =$
155,800

(A) Square reflecting the denominator in the correlation formula used in the sample problem

$\Sigma xy = 370$

$(\Sigma xy)^2 = 136,900$

(B) Square reflecting the numerator in the correlation formula used in the sample problem

$\sqrt{\Sigma x^2 \cdot \Sigma y^2}$ Σxy

(C) Ratio between the sides of the two squares which defines the correlation coefficient

FIGURE 11.1 Graphic portrayal of correlation coefficient

Similarly, the numerator in the r formula incorporates deviations found in the two distributions. The numerator (that is, Σxy) represents one side of a square that would characterize the actual amount of squared deviations the two distributions have in common. This is shown graphically in part B of Figure 11.1.

What we have, then, in the r formula is a ratio between how much score deviation the two distributions actually have in common (that is,

Σxy) and the $\underline{\text{maximum}}$ amount of score deviation that they could have in common ($\sqrt{\Sigma x^2 \cdot \Sigma y^2}$). Or in terms of the graphic portrayals in Figure 11.1, r is the ratio of the side of the smaller square to the side of the larger square, as shown in part C.

We can also consider correlation from the standpoint of the amount of actual variance that the two distributions have in common. This is equal to r^2, or, in terms of part C in Figure 11.1, it is the ratio of the area in the smaller square to the area in the larger square.

The Significance of r

Under the terms of the null hypothesis, we can ask what the probability would be of obtaining an r of the magnitude of the one found. One method of estimating this probability is with the use of the t distribution. Using the procedure shown in Table 11.1, the value obtained for r is entered into a special formula for t. Given a value of t, it is interpreted for an associated probability level by consulting a sampling distribution of t (usually in tabular form such as Table 7.2). The degrees of freedom for entering the table are the number of pairs of scores, minus 2 ($N - 2$). Given the present example, with $t = 7.59$ and with $d.f. = 8$, and referring to Table 7.2, we see that p falls far below the .01 level. This tells us that with a null hypothesis of $r = 0$, we would expect to find an r of .93 in less than one out of 100 cases if it were due only to sampling error.

The fact that the correlation is statistically significant, however, does not mean that it is necessarily meaningful to us. What must be done next is to interpret its magnitude and direction.

The Interpretation of r

We have seen that there are three things to consider when given a particular value of r. First, there is the magnitude of the coefficient—it may vary from a zero or negligible level to a level that may approach perfect correlation (1.0). Second, there is the direction indicated by the sign of the coefficient. Do the measures vary directly (positively), or do they vary inversely (negatively)? Third, as discussed in the preceding section, there is the question of whether the obtained r is significant. That is, under the null hypothesis of $r = 0$, what is the probability of obtaining the value of r that was actually found? If this probability is sufficiently low (that is, below the level set for rejection), we would reject the null hypothesis in favor of the research hypothesis that the two measures are indeed correlated.

In most practical situations, we would probably consider the foregoing three points in just their reverse order. That is, before considering the direction and magnitude of a coefficient, we would first want to know whether it is significant or not. If we are not in a position to reject the null hypothesis of $r = 0$, we would simply report this outcome as the sole interpretation of r. There would be no particular utility in discussing direction and magnitude of an r that perhaps represents only sampling error.

Given an r that is significant, its interpretation depends substantially on the nature of what is being studied. The direction of the coefficient, of course, is expected to conform with whatever expectations the researcher holds. If such expectation is fulfilled, the interpretation of direction is a straightforward matter.

By contrast, the interpretation of the magnitude of a correlation is often a subjective matter. Practically speaking, a relationship that might be considered moderate in one situation might be considered relatively negligible in another. Here we are talking about the psychological implication of the coefficient, not simply its magnitude in the statistical sense. For example, we might be interested in knowing what factors are related to a person's ability to write news stories. Knowing that, for instance, a person's grades in college English composition classes are correlated at the level of about $r = .42$ with ratings of news-writing ability seems of moderate consequence. The magnitude, although low in a statistical sense, is of sufficient psychological consequence that we might take this factor, among others, into account when considering what is related to news-writing ability.

But on the other hand, suppose that our purpose were to have a good prediction index of news-writing ability. The correlation of .42 is relatively negligible for this task. We would probably look for other factors as a basis for prediction. Thus, even though a correlation of .42 might be significant statistically, it may still be of relatively little consequence in terms of the problem under study. This last point perhaps needs special emphasis: Even though a correlation is statistically significant, its psychological significance remains to be interpreted by the researcher. In other words, statistical significance only puts researchers in a position to interpret the magnitude of the correlation; it does not do this job for them.

Even though the verbal description of a correlation remains highly contingent on what is being assessed, it is still useful to have some consistency of terminology in describing the magnitude of the coefficient itself. Although there is not a great deal of consistency in the research literature, Guilford[1] has suggested the following as a rough guide:

$<.20$ slight; almost negligible relationship
$.20-.40$ low correlation; definite but small relationship
$.40-.70$ moderate correlation; substantial relationship
$.70-.90$ high correlation; marked relationship
$>.90$ very high correlation; very dependable relationship

---

[1] J.P. Guilford, *Fundamental Statistics in Psychology and Education* (New York: McGraw-Hill, 1956), p. 145.

The Nature of r as an Index

One final point is a warning that a correlation coefficient says nothing about a percentage of relationship. That is, a correlation of .40 is not twice the relationship of one of .20. The coefficient itself is simply a convenient index, not an actual measurement scale.

In order to gain some feel for what an r implies, it is often useful to consider what r^2 signifies. As discussed earlier, r^2 is the proportion of variance that two measures have in common. In contrast to r, the value of r^2 can be thought of in terms of a proportion of percentage of relationship. When used for this purpose, r^2 is called a *coefficient of determination.* In terms of a coefficient of determination, we can say that two measures with an r^2 of .50 have twice the variance in common as two measures where r^2 = .25.

Figure 11.2 illustrates what is meant by having a feel for what r implies in terms of the statistical relation between two measures. Consider this graph of the relationship between r and r^2. Notice that relatively low

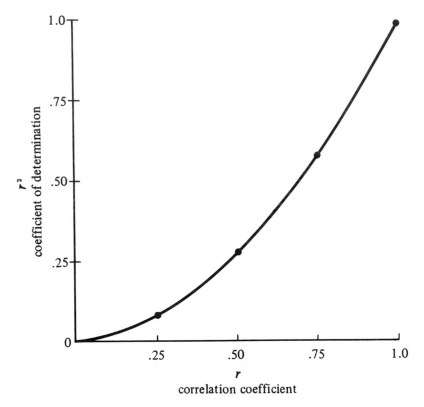

FIGURE 11.2   The relationship between correlation coefficients (r) and coefficients of determination (r^2)

values of r (for instance, .25) indicate an almost negligible amount (for example, $r^2 = .06$) of common variance between two measures. By contrast, note that as r becomes relatively large (for instance, .90) the corresponding proportion of common variance ($r^2 = .81$) is closer to the value of r; there is a substantial overlap in the variance of the two measures.

Another useful way to depict what r implies is shown by the overlap among circles as illustrated in Figure 11.3. Note that for the varying values of r, we can see r^2, or common variance, in terms of the overlap between the circles.

Not only is the coefficient of determination, or r^2, helpful in considering magnitude of correlation, it becomes particularly useful when we are

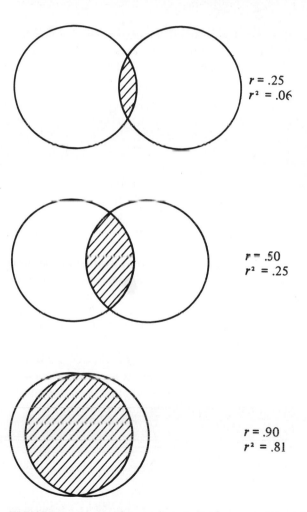

$r = .25$
$r^2 = .06$

$r = .50$
$r^2 = .25$

$r = .90$
$r^2 = .81$

FIGURE 11.3   Graphic portrayal of common variance (shaded area) associated with different values of r

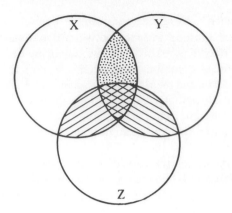

FIGURE 11.4   Graphic portrayal of a three-variable problem

considering correlation with reference to predicting values of one measurement from one or more other measurements, a topic treated in Chapter 12. Furthermore, the idea of common variance among measures is a central concept in the logic of factor analysis, the topic of Chapter 15.

MULTIPLE CORRELATION

Thus far, we have considered correlation where only two variables have been involved. Consider the case where we have three variables, X, Y, and Z, as shown in Figure 11.4. One way to describe the relations among X, Y, and Z is to consider them in terms of a series of two-variable problems. That is, we could calculate r_{XY}, r_{XZ}, and r_{YZ}, or r^2_{XY}, r^2_{XZ}, and r^2_{YZ} and, thus, account for the relations shown in the three overlapping circles. We would probably report the results of the r calculations in the form of a table called an *intercorrelation matrix*, such as is presented in Table 11.2. Here we can specify the correlation of each of the three variables with each other. Obviously, such a matrix could be extended to accommodate the intercorrelations among any number of variables.

Rather than assessing the interrelations among more than two variables as a series of two-variable problems, it is possible to assess the degree to which two or more variables together relate to a given variable. This is a problem in what is called *multiple correlation*.

Table 11.2   **Example of a Matrix of Intercorrelations for a Three-Variable Problem**

	Y	Z
X	.50	.40
Y	—	.45

Coefficient of Multiple Correlation

As an index of the relation of two or more variables with a given variable, the *coefficient of multiple correlation* is symbolized by R. The coefficient R has the same scale as the two-variable coefficient, r, that is,

$$+1.0 \text{ (perfect positive correlation)}$$
$$0.0 \text{ (no correlation)}$$
$$-1.0 \text{ (perfect negative correlation)}$$

The R symbol usually carries subscripts that indicate which variables are seen as related to a given variable. Suppose that Z is the variable focused upon, and R indexes the multiple correlation of X and Y with Z. The subscripts would be as follows:

$$R_{Z.XY}$$

whereas $R_{X.YZ}$ would symbolize the multiple correlation of Y and Z with X, or $R_{Y.XZ}$ would symbolize the multiple correlation of X and Z with Y. Similar to r^2, we can consider R^2 as a coefficient of multiple determination. The common variance defined by $R^2_{Z.XY}$ can be seen as the shaded area in Figure 11.4, that is, the area that Variables X and Y have in common with Variable Z. (You might consider also how $R^2_{X.YZ}$ and $R^2_{Y.XZ}$ would be shown in Fig. 11.4.)

A SAMPLE STUDY

Consider a study where the researcher wants to assess the relationship between how individuals rate television programming (Variable X) and the price paid for television sets (Variable Y), with the average number of hours that they spend watching television during a week (Variable Z). This defines a three-variable problem that will center on the calculation of $R_{Z.XY}$. Suppose that the number of individuals (N) surveyed equals 100.

The necessary data for calculating R are given in Table 11.3. This body of data comprises what was calculated (calculations not shown) as the individual correlation coefficients among the three variables—r_{XY}, r_{YZ}, and r_{XZ}. Using the formula given in Table 11.3, $R_{Z.XY}$ is calculated. In rounded form, it equals .97.

A coefficient of multiple correlation is interpreted in much the same manner as was r in a two-variable problem. Although we shall not go into the details of calculation, an obtained value of R can be tested against the null hypothesis that $R = 0$. This yields a value that can be interpreted to tell us the probability that the obtained multiple R could occur by sampling error. If this probability value is below that set for rejection, we are in a position to accept the research hypothesis that the Variables X and Y have a significant multiple correlation with the Variable Z. If the

Table 11.3   **Example Calculation of a Coefficient of Multiple Correlation (Simulated Data)**

Given:

$$r_{XY} = .750$$
$$r_{YZ} = .918 \qquad N = 100$$
$$r_{XZ} = .894$$

Solution:

$$R^2_{Z.XY} = \frac{r^2_{XZ} + r^2_{YZ} - (2r_{XZ}r_{YZ}r_{XY})}{1 - r^2_{XY}}$$

$$= \frac{(.894)^2 + (.918)^2 - 2(.894)(.918)(.750)}{1 - (.750)^2}$$

$$= \frac{(.799) + (.843) - (1.231)}{1 - (.563)}$$

$$= \frac{.411}{.437} = .9405$$

$$R_{Z.XY} = \sqrt{.9405} = +.9698$$

multiple correlation coefficient (R) is significant, then we would interpret its direction and magnitude much the same as was the case with r.

BINARY OR "DUMMY" VARIABLES

It is possible to estimate the correlation of continuous variables (for example, interval scale) with categorical (nominal scale) ones by using *binary* (or *dummy*) variables where classes are coded by uses of "0" and "1." Suppose, for example, that in the earlier sample study (Table 11.1), in addition to the variables of reading aptitude (X) and hours spent reading books (Y), there was an additional variable of respondent sex (Z). Also, let us assume that the first five Ss were male, so their score of Variable Z is each "0," and the second five are female, so their score of Variable Z is each "1" (or the numbers for the sexes could be reversed).

Using the formula for two-variable correlation, we would find that, as before (Table 11.1)[2]:

$$r_{XY} = .937$$

The same formula would also indicate that:

$$r_{YZ} = .351$$
$$r_{XZ} = .387$$

[2] There are more precise methods for calculating correlations involving binary variables, but the present formula suffices for this illustration.

Sometimes the dummy-variable approach is useful when we want to index the relationship of a group or treatment variable with a dependent or differentiating measure. Recall, for example, in Chapter 7 the example of two groups of children where one (I) had seen a cartoon showing aggressive activities, the other (II) saw a nonaggressive one, and the dependent variable was a measure of aggressiveness of the children's play activities after seeing one or the other cartoon. In the calculational example for t (Table 7.1), it was shown that the difference in means on the aggressiveness measure was statistically significant ($.02 < p < .05$). Thus, it was concluded that seeing an aggressive cartoon as compared with a nonaggressive one was related to aggressiveness in the children's behavior. Given this outcome, we would expect to see a correlation between group membership (coded "0," "1") and the aggressiveness measure. Given Group I coded as "0" and Group II as "1," the correlation of this coded group variable with the aggressiveness measure is:

$$r = -.66$$

If more than two categories must be coded, *each* can be included as a separate dummy variable. For example, to extend the above example, as in Chapter 8, there could be three groups: one seeing a cartoon where aggression occurs but punishment follows, another seeing unpunished aggression, and a third seeing a cartoon with no aggression. A child's membership in *each* group could be coded as "0" (not in) or "1" (in).[3]

PARTIAL CORRELATION

It is also possible to describe the correlation of one variable with another while ruling out that degree of correlation attributable to other variables. Consider, for example, the three-variable graphic portrayal of correlation in Figure 11.4. The overall correlation of X and Z would be the total that the two circles overlapped. The correlation that both X and Z have in common with Y is the center cross-hatched area. If this were removed from the overlap between X and Z, the remaining section (diagonal lines) would represent the variance involved in the partial correlation ($r_{XZ.Y}$) of X with Z, with the joint effects of Y nullified.

For a numerical example, consider the intercorrelations summarized in Table 11.3 for the following variables:

X = ratings of television programming
Y = price paid for television set
Z = time spent watching television each week
$$r_{XY} = .750 \quad r_{YZ} = .918 \quad r_{XZ} = .894$$

---

[3] Actually, only two of the three groups need be coded to do the analysis because lack of membership in two implies membership in a third.

Table 11.4   Example Calculation of a Coefficient of Partial Correlation

Given:

$$r_{XY} = .750$$
$$r_{YZ} = .918$$
$$r_{XZ} = .894$$

Solution:

$$r_{XZ.Y} = \frac{r_{XZ} - r_{XY} r_{ZY}}{\sqrt{1 - r^2_{XY}} \sqrt{1 - r^2_{ZY}}}$$

$$r_{XZ.Y} = \frac{.894 - (.75)(.918)}{\sqrt{1 - .75^2} \sqrt{1 - .918^2}}$$

$$r_{XZ.Y} = .783$$

Suppose, then, that we wished to know the correlation of television ratings (X) with time spent watching television (Z) while controlling for price of set (Y). Calculations are summarized in Table 11.4:

$$r_{XZ.Y} = .783$$

The result indicates a slight drop from the original correlation (.894), indicating that much of the XZ relationship is independent of joint relations with Y.

NONPARAMETRIC APPROACHES TO CORRELATION

The most common nonparametric alternative to Pearson's product-moment coefficient is a method known as *Spearman's rho*, sometimes called *rank-difference* correlation. This method is subject to less error than the product-moment formula when samples are relatively small (say, <30) and also when measurement has only the power of an ordinal scale.

Table 11.5 illustrates how rho (ρ) would be calculated for the data presented in the sample two-variable study. Note first that the actual values entered into the formula are based on rankings of the two sets of values. For each pair of ranking values x_r, y_r, a difference (D) value is obtained. These difference values are squared, then summed and entered into the formula for ρ. Generally, ρ is interpreted in much the same way as Pearson's r.

Another type of rank-order correlation method is *Kendall's tau*. Although its calculation is different from rho, the tau coefficient has many similar characteristics and is interpreted accordingly.

Table 11.5   Example of the *Spearman Rho* Correlation (Skeletal Data from Table 11.1)

Ss	Raw Data		Ranking Values			
	X (Reading Aptitude)	Y (Hours)	x_r	y_r	D $(x_r - y_r)$	D^2
1	20	5	6	5.5	0.5	0.25
2	5	1	2.5	1	1.5	2.25
3	5	2	2.5	2.5	0.0	0.00
4	40	7	9.5	7	2.5	6.25
5	30	8	7	8.5	−1.5	2.25
6	35	9	8	10	−2.0	4.00
7	5	3	2.5	4	−1.5	2.25
8	5	2	2.5	2.5	0.0	0.00
9	15	5	5	5.5	−0.5	0.25
10	40	8	9.5	8.5	1.0	1.00
						Σ 18.50

Formula:

$$\rho = \frac{6\Sigma D^2}{n(n^2 - 1)}$$

$$= 1 - \frac{6(18.50)}{10(100 - 1)}$$

$$= 1 - .112$$

$$\rho = .888 \text{ (or .89)}$$

SUMMARY

Correlation characterizes the relationship between variables—that is, the degree to which two variables vary together (*positive* correlation) or vary inversely (*negative* correlation). The *Pearson product-moment* correlation coefficient, r, has a range of values from +1.0 (perfect positive correlation) through −1.0 (perfect negative correlation). In addition to the direction and magnitude of r, we can also test the null hypothesis of r = 0 against a research hypothesis of r ≠ 0.

The calculation of r reflects a ratio between the maximal amount of variability that two measures could have in common and the amount that they actually have in common. The coefficient squared (r^2) is the actual proportion of variance that two measures have in common.

A coefficient of multiple correlation (R) indexes the degree to which two or more variables are related with a given variable. The coefficient R is interpreted similarly to r.

Correlation with a categorical variable can be calculated where dummy or binary ("0," "1") coding represents the categories.

Partial correlation provides a distinction of a relationship between two variables while controlling for effects of a third variable.

Selected nonparametric correlational models include *Spearman's rho* and *Kendall's tau*.

SUPPLEMENTARY READINGS

Hays, William L., *Statistics*. 3rd ed. New York: Holt, Rinehart and Winston, 1981. Chapter 13 introduces correlation and regression.

Howell, David C., *Fundamental Statistics for the Behavioral Sciences*. Boston: PWS-Kent, 1989. See Chapter 9.

SAS Institute, Inc., *SAS User's Guide: Statistics*. 1982 ed. Cary, N.C.: SAS Institute, 1982.

SPSS, Inc., *SPSS-X User's Guide*. 3rd ed. Chicago: SPSS, 1988. Chapter 24 describes a computer program for correlation, and Chapter 37 describes a program for calculating nonparametric correlation analysis.

12

Regression

In the previous chapter, we were concerned mainly with the relation, or "correlation," between or among variables. The correlation coefficient, for example, was seen as simply indexing the degree to which two or more measures varied together. Or the coefficient of determination was seen as indexing the amount of variance that two sets of measurements had in common. Obviously, knowledge of the relation among variables can be more than an end in itself. Sometimes we want to take advantage of a relationship and use it as a basis for prediction. That is, given a knowledge of the variable X and its relationship with the variable Y, how can we use particular values of X to predict corresponding values of Y? The main statistical tool for answering this question is regression analysis.

A SAMPLE STUDY

For purposes of illustration, let us consider the two-variable problems presented in the preceding chapter where Variable Y was the average number of hours a person spent reading books during the week and Variable X was a person's score on a reading aptitude test. When using regression analysis, we are essentially interested in the description of a predictive relationship. Let us assume that we want to predict persons' times spent in reading books on the basis of knowing their reading aptitudes. We can calculate this prediction if we have an array of scores for the two variables or even if we simply have their means, the standard errors of those two means, and the correlation between the two variables.

Problem

Given that we know from the sample study in Chapter 11 (Table 11.1) that a correlation of .937 has been found between measures of reading aptitude and the average number of hours that people spend in reading books, we are interested in calculating a formula that will provide us with predicted values of the latter given the knowledge of values of the former. We are also interested in the reliability of our estimates.

Method

Taking the same scores from the sample of 10 adults described in Table 11.1 and using a statistical method called *linear regression*, the following results are calculated (these calculations are subsequently described in Table 12.1):

$$Y' = .181X + 1.39$$

Table 12.1   **Example Calculation of the Regression Equation for the Prediction of Y Values from X Values (Simulated Data from Table 11.1)**

Composite formula:

$$Y' = r_{XY}\left(\frac{s_Y}{s_X}\right)(X - M_X) + M_Y$$

where Y' is the predicted value of the Y variable
r_{XY} is the correlation between X and Y ($r_{XY} = +.937$)
s_X and s_Y are the standard deviations of the X
and Y distributions ($s_X = 14.32$; $s_Y = 2.76$)
X is a given value of the X variable
M_X and M_Y are the sample means of the distributions
of the X and Y variables ($M_X = 20$; $M_Y = 5$)

Calculations:

$$Y' = .937\left(\frac{2.76}{14.32}\right)(X - 20) + 5$$
$$= (.181)(X - 20) + 5$$
$$= (.181X) - 3.61 + 5$$
$$Y' = .181X + 1.39$$

In this formula, Y' is the predicted number of hours that a person will spend reading books during each week of the year, whereas X is the score for that person on the reading aptitude test. With this formula, when a given value of X is multiplied by .181 and then added to 1.39, we have our predicted value of Y. Suppose, for example, that for a given individual, the reading score (X) is 40. As applied in this formula, the results would represent a prediction (Y') of 8.63 hours per week. Or a reading aptitude score of 15 would predict 4.105 hours of reading per week, and so on.

How good is this estimate? Using an appropriate formula (discussed later) we can calculate what is known as *standard error of the estimate* (SEE). This will tell us that we can expect a particular amount of variation in our prediction given a particular probability. For example, we could conclude that the probability is 68 percent that our estimate will be within $+.96$ or $-.96$ of the value of Y that we calculated.[1] Or we could say that for a probability of 80 percent we could expect our estimate not to vary outside of $+1.23$ or -1.23. From these two examples, you can see that the larger the range of allowance in our estimate, the greater is the probability that the true prediction is *within* that range, or in terms of error, the less is the probability that the true prediction is *outside* that range.

[1] This SEE is relatively large, in part because of the small size of sample used for illustrative purposes.

We can also test the statistical significance of the regression equation by use of an F test. In this case (calculation given later), $F = 57.33$, which we can interpret as statistically significant $(p < .01)$.

Results

A practical conclusion to this analysis, stated in mainly verbal terms, might be as follows:

The number of hours that a person spends in reading books each week can be quite accurately predicted from a measure of reading aptitude. A statistically significant prediction equation was

Reading time = (.181 × reading scores) + 1.39

With 20% probability of error, we can assume that our estimates will be accurate within +1.23 or −1.23 of the actual estimated figure.

THE LOGIC OF REGRESSION ANALYSIS

The term "regression" does not help much in introducing one to the logic of this type of prediction analysis. It is more useful to consider how a line on a graph can describe the relationship between sets of scores on two variables. This is called a *regression line* or *line of best fit*.

Line of Best Fit

Suppose that we have two variables, P and Q, and that we have four pairs of such scores.

	(P)	(Q)
Pair 1	10	7
Pair 2	20	12
Pair 3	30	17
Pair 4	40	22

We have plotted these scores in the graph shown in Figure 12.1; each pair is located in terms of the intersection of its P and Q scores on the graph. Notice that if we connect these points with a straight line, we have a basis for predicting further corresponding pairs of P and Q scores. Thus, for example, if the relation between P and Q is, indeed, what we have observed, then if $P = 15$, $Q = 8.5$, if $P = 25$, $Q - 13.5$, or if $P = 38$, $Q = 20$, and so on.

The line in Figure 12.1 is a line of best fit or regression line. It defines our basis for predicting values of Q, given values of P (and vice versa in this particular example).

Suppose that we now plotted the pairs of scores from the sample study on a similar graph, as in Figure 12.2. Here we have the scores for each of the 10 Ss plotted in terms of the intersection of each pair of X and Y values. Notice first that we cannot connect these points with the same

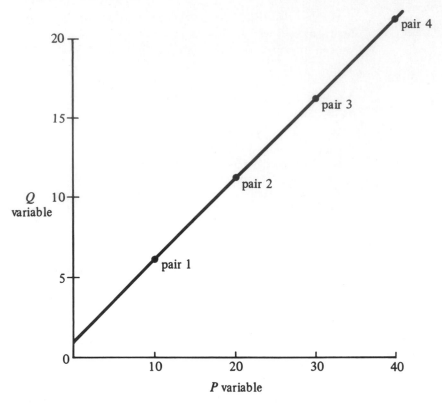

FIGURE 12.1   Distribution of pairs of P and Q scores and line of best fit

kind of straight line as in the previous example. This is because the relation between X and Y values is not perfectly consistent as it was with the pairs of P and Q scores. In other words, although we know that X and Y values tend to vary together (from their correlation calculated in Chapter 11), we do not have a perfect positive relationship ($r_{XY} = .94$, whereas $r_{PQ} = 1.0$).

Even though the relation between X and Y is not perfect, we can still use the concept of the line of best fit. This time, however, the regression line is located such that given the observed relation between X and Y, we can attempt to make predictions of Y which, although not perfect, would involve the least degree of prediction error. In other words, our regression line is now a compromise in getting the line of best fit. This line then serves as our basis for predicting Y values, given various values of X.

Our computational procedure for obtaining this line will be discussed at a later point. The present line applies only for predicting Y values, given X values, and not the reverse. (The reverse would involve location

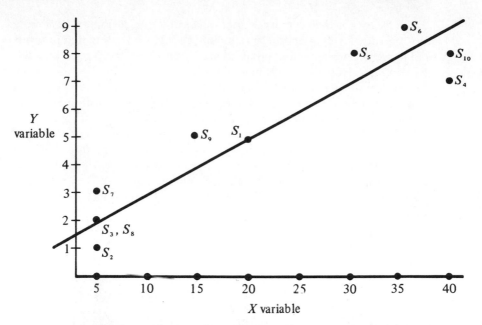

FIGURE 12.2   Distribution of X and Y scores from the sample study, including a line of best fit for the regression of Y on X (calculations in Table 12.1)

of a slightly different line, one giving the best estimate of X values, given various values of Y.) The present approach also assumes that a straight line will be the best basis for prediction, a procedure known as *linear regression*. There are alternative procedures for problems involving non-linear regression (see Supplementary Readings at the end of the present chapter).

Regression Equation

We usually describe the line of best fit in terms of a mathematical formula called a regression equation. This equation tells us the best-fitting mathematical linear relation between X and Y values. That is, given a value of X, it tells us what must be done to derive the best estimate of a value of Y. This equation is actually the mathematical expression for a straight line, an equation that you may have been introduced to in elementary algebra. Its general form is:

$$Y = bX + a$$

For a given line, *a* defines what is called its *intercept*, whereas *b* defines its *slope*. Notice in Figure 12.1 that as the line approaches the smaller values of P and Q, it intersects the Q axis at 2.0. This means that in terms of intercept (and we must still consider slope), the variable Q has values

2.0 units greater than corresponding values of P. Or, in Figure 12.2, note that the intersection on the Y axis is at about 1.39, indicating that in terms of intercept, corresponding points along Y exceed points along X by 1.39. By contrast, slope (or b) tells how many units the variable Y is increasing for every unit increase in X. For example, in Figure 12.1, b is equal to 0.5, which means that (in addition to intercept) for every unit increase of P, there is a 0.5-unit increase in Q. Similarly, in the case of Figure 12.2, b is equal to .181, indicating that for every unit increase in X, Y increases .181 units. Imagine what would happen to the slope of the line if b were increased. It would become steeper when an increase in a unit of X would have a greater increase in the corresponding unit of Y. Or if b were decreased, increases in a unit of X would be associated with lesser increases in a unit of Y.

Generally speaking, a and b are usually referred to as *regression co-efficients* (or sometimes *constants*). Given the knowledge of what a and b are equal to in a given problem, we can then use the above equation to estimate a value of Y, given a particular value of X. (We indicate that Y is an estimated value by giving it a prime sign, that is, Y', or when estimating Q, we write Q'.)

For some practical examples in the use of this equation, consider first the perfect relation between P and Q shown in Figure 12.1. Given that $a = 2.0$ and $b = 0.5$, then:

$$\text{If } P = 10, \text{ then } Q' = (.5 \cdot 10) + 2 = 7, \text{ or}$$
$$\text{If } P = 20, \text{ then } Q' = (.5 \cdot 20) + 2 = 12, \text{ and so on}$$

Similarly, for the line of best fit shown in Figure 12.2, if $a = 1.38$ and $b = .181$, then:

$$\text{If } X = 10, \text{ then } Y' = (.181 \cdot 10) + 1.39 = 3.2, \text{ or}$$
$$\text{If } X = 40, \text{ then } Y' = (.181 \cdot 40) + 1.39 = 8.63, \text{ and so on}$$

In order to see again the relation between the above equation and the line of best fit, note how the estimated values of Y (Y') and values of Q' fall on their respective lines of best fit (Figs. 12.1, 12.2) at the intersection of appropriate X and Y values (or P and Q values).

Calculation of a and b

The essence of the calculation in the foregoing formula is to determine what a and b are equal to for a given problem. Table 12.1 provides an appropriate formula and illustrates how this formula would be used for calculations applicable to the sample study where X is a person's score on a reading aptitude test, and Y is the average number of hours a person would spend in reading books during a week. Notice that the equation appears in a composite form; that is, a and b are both calculated with the single formula shown in Table 12.1. Generally, what we have described as b corresponds to the $r_{XY}\left(\dfrac{s_Y}{s_X}\right)$ portion of the formula, and $(X - M_X)$ + M_Y corresponds to a. As shown in Table 12.1, the solution gives us the following equation:

$$Y' = .181X + 1.39$$

which is the same as discussed earlier and which describes the line of best fit shown in Figure 12.2. The foregoing calculations were our basis for determining this line.

In practical terms, again, we could say that:

$$\text{Reading time} = (.181 \times \text{reading scores}) + 1.39$$

We would use the above equation in order to make whatever predictions we desired, or we might construct a graph such as shown in Figure 12.2.

Standard Error of the Estimate

As described earlier in the sample study, the *standard error of the estimate* (SEE) provides an estimate of the accuracy of our prediction (or error, depending on how we look at it). Note in Figure 12.2 how the actual values of Y scores ($S_1, S_2, \ldots, S_{10}$) vary about the regression line. As you would logically expect, the greater that these values vary from those potentially based on the regression equation (that is, those on the regression line), the greater is the probability of error in making predictions from that equation (or line). Error increases with the decrease in linear correlation between the two variables. Note the typical equation below for calculation of the standard error of the estimate (SEE) for the prediction on Y from X:

$$SEE_{YX} = s_X\sqrt{1 - r^2_{YX}}$$

If you examine the components of this equation, you will see that the greater the correlation of X and Y (r_{YX}), the less will be SEE. You may also note that the greater the variability or standard deviation of the X distribution (s_X), the greater will be the standard error of the estimate.

As implied by its label, SEE is interpretable as a standard error score on a normal curve. More practically, the calculation tells us what an error score (of *one*) is equal to in terms of the Y scale of scores. It is, thus, interpretable in terms of probability values. Recall in Chapter 5 how *one* standard error of the mean encompassed 34 percent of a normal distribution on each side of a mean, or 68 percent, counting both. In the case of SEE, we see what one standard score is equal to in our predicted score values; then, by interpreting a normal curve table, we can make such statements that if SEE = .963 we would expect the true prediction of *Y* to fall inside plus or minus 1 SEE of the estimate of *Y* 68 percent of the time, or 32 percent outside of it.

A simple way to visualize standard error of the estimate is to plot it along with a regression line, and this has been done in Figure 12.3 for the sample study. The regression line is the same as in Figure 12.2 but now we have added a dashed line above and below the regression line to represent the 34 percent probability that estimates of *Y* may be higher than predicted by the regression line and the 34 percent probability that they will be less. Figure 12.3 is a useful reminder that in using regression equations we are making estimates, all with calculable margins of error.

Application of the *F* Test

There are several strategies for testing the statistical significance of a simple regression analysis. One way is to test the significance of the correlation coefficient such as described earlier in Chapter 11, p. 136.

Another, and more often used, method is by use of an *F* test, an approach not unlike that described in Chapters 8 and 9 on analysis of variance (see especially pp. 93–95). In fact, one reason it is useful to understand how *F* is used in regression is that it reveals how regression and analysis of

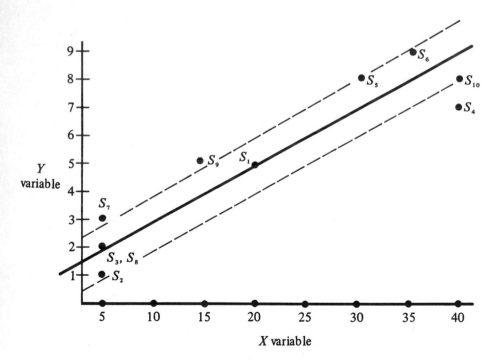

FIGURE 12.3   Range of deviation of *Y* scores one SEE above and one SEE below the line of best fit for sample study and where SEE = .963

variance are based on the same underlying concept of dividing total variance into components. There is variance that is "accounted for" (such as *between groups* in analysis of variance) and variance that is "unaccounted for" (such as *within groups*) in the sense that it is attributable to sampling variability. Recall (p. 94) that an *F* ratio was:

$$F = \frac{\text{variance between groups} \div \text{degrees of freedom}}{\text{variance within groups} \div \text{degrees of freedom}}$$

Given a calculated value of *F* and the degrees of freedom associated with the components of the *F* ratio, we can then make probability estimates that the *F* occurred by chance. We can test hypotheses with this *F*.

In regression analysis, *F* essentially represents:

$$F = \frac{\text{variance predicted (or "regression")} \div \text{degrees of freedom}}{\text{variance unpredicted (or "residual")} \div \text{degrees of freedom}}$$

Logically, the more that the predicted variance exceeds the unpredicted variance, the less we would expect the value of *F* to have arisen by chance.

The components for *F*, as in analysis of variance, are drawn from sums of squares. The numerator, or predicted variance, is the sum of squared deviations of the predicted variable about its mean, or simply the sum of squares due to the "regression." The denominator is the sum of squared deviations of differences between the known values of the predicted variable and the predicted values or is simply the sum of squares of the "residual." (Total variance would be the sum of squared deviations of the known values of the predicted variable.)

For the sample study of the prediction of hours spent in reading books (*Y*) from scores on a reading aptitude test (*X*), *F* = 57.9. The degrees of freedom for this *F* are the number of variables minus one, and the size of sample minus the number of variables minus one. *F* is interpreted in a table of *F* (as in the Appendix). Here we find the *F* of 57.9 to be statistically significant ($p < .01$). We could, thus, reject a null hypothesis that our predictions were no better than chance.

F tests are particularly useful in regression equations with more than one predictor variable ("multiple regression," Chapter 13). We can, for example, test whether each additional predictor variable contributes significantly when added to the equation.

FURTHER NOTES ON REGRESSION

Use of Binary or Dummy Variables

In Chapter 11 (pp. 142–143), we illustrated how categorical variables could be entered in a correlation analysis by using "0" and "1" scores to code group or treatment membership. Recall how data from the sample study in Chapter 7 (Table 7.1) were used to calculate the correlation of a measure of a child's aggressive play with whether the child had just seen a cartoon depicting aggressive behavior (Group I, coded "0") or one not showing aggression (Group II, coded "1"). The correlation was $r = -.66$, indicating a relationship between exposure to cartoons depicting aggression and aggression in play.

Similarly, binary or dummy variables can be entered as predictors (only) in a regression equation. Using the data from Table 7.1 and coding the groups as described above as the Y variable and scores on aggressive behavior as the X variable, the resulting equation is:

$$X' = -5Y + 57$$

Here we would not be especially interested in using the formula as a basis for predicting values of the aggressive play scores (X) on the basis of treatment groups membership (Y). (Note how respective binary values of Y, 0 or 1, result in predicting the means of the two groups, 57 or 52.) Instead, we could take our basis for calculating this equation and calculate an F ratio, which in this case is 6.25. At 1 and 8 $d.f.$, this is statistically significant $(p < .05)$. In brief, F can be calculated in this way to test the statistical significance of a hypothesis of group (or treatment) differences.

Dummy variables can be used as above in the role of a substitute for a t test or analysis of variance of group differences. However, you will find them more often in communication and education studies where the researcher wishes to test the role of a nominal variable along with other variables in a problem of multiple prediction (Chapter 13).

Relationship with t Tests and Analysis of Variance

Several times, we have noted that regression analysis and analysis of variance are generally two ways of looking at the same thing. They involve dividing total variance into "accounted for" components and estimating the probability that the variance accounted for is greater than could occur by chance alone.

If we performed an analysis of variance on the data in Table 7.1, it, like the binary regression analysis, would result in an $F = 6.25$. Similarly, to recall the relation between t and F $(F = t^2)$ mentioned briefly in Chapter 8, the t of 2.50 reported in Table 7.1 also equals an F of 6.25. In short, t, F, and linear regression analysis are all very similar in their underlying components.

SUMMARY

In addition to indexing the relationship between variables, a regression equation can also serve as a basis for *prediction*. That is, given values of one variable, what are the best estimates of related values of another variable? The basis for such prediction is found in the logic of regression analysis. The predictive relationship can be shown in terms of a *line of best fit,* or in terms of the mathematical definition of this line, called a *regression line.* We can calculate a range of probable error for our predictions, that is, the *standard error of the estimate* (SEE). Also, we can test whether predicted variance exceeds a chance occurrence by use of the *F* test (the same essentially as in analysis of variance).

SUPPLEMENTARY READINGS

Howell, David C., *Fundamental Statistics for the Behavioral Sciences.* Boston: PWS-Kent, 1989. See Chapter 10.

SAS Institute, Inc., *SAS User's Guide: Statistics. 1982* ed. Cary, N.C.: SAS Institute, 1982. See the section on "Regression."

SPSS, Inc., *SPSS-X User's Guide.* 3rd ed. Chicago: SPSS, 1988. Chapter 45 describes a program for regression analysis.

13

Multiple Regression

Just as we can assess the multiple correlation of two or more variables with another variable, it is also possible to bring more than one predictor variable to bear in predicting scores on a given variable. This technique is known as multiple regression analysis. *It can be applied for practical purposes of predicting a score on a variable given scores on two or more predictor variables. It is also used as a method of describing the relative degree of contribution of a series of variables in the multiple prediction of a variable.*

A SAMPLE STUDY

Problem

Recall in Chapter 11 the example where it was determined that the multiple correlation of a person's evaluation of television programming (Variable X) and the price paid for the television set (Variable Y) was $+.97$ (rounded) with the number of hours a person would spend watching television per week (Variable Z). In the present sample study of multiple regression, we are interested in achieving four objectives: (1) developing a formula for practical prediction of Variable Z from Variables X and Y, (2) determining whether the predictions are better than chance, (3) estimating the accuracy of the predictions, and (4) describing the relative contributions of the Variables X and Y to the prediction of Variable Z.

Method

Using a calculation procedure (described subsequently in Table 13.1), we determine that the prediction formula is as follows:

$$Z' = (-10.13) + 2.51X + .076Y$$

Here Z' is the variable that we predicted, or the number of hours that a person would spend watching television per week. The Variable X is whatever scaled rating we have of a person's evaluation of a television program, and Variable Y is the price the person paid for the television set. Further calculations are made for F and standard error of the estimate.

Results

Given the foregoing formula as a basis, the results can be summarized as follows:

1. Viewing time $= (-10.13) + (2.51 \times$ programming rating$) + (.076 \times$ price of set$)$.

Table 13.1 **Example Solution to Equation of the Multiple Regression of Z on X and Y (Simulated Data from Table 11.3)**

Given:

$$r_{XY} = .750 \qquad M_X = 5 \qquad s_X = 2.2$$
$$r_{YZ} = .918 \qquad M_Y = 205 \qquad s_Y = 88.0$$
$$r_{XZ} = .894 \qquad M_Z = 18 \qquad s_Z = 11.8$$
$$N = 100 \text{ cases}$$

Beta (β) coefficients:

$$\beta_{XZ.Y} = \frac{r_{XZ} - r_{YZ}r_{XY}}{1 - r_{XY}^2}$$

$$= \frac{(.894) - (.918)(.750)}{1 - (.750)^2}$$

$$= \frac{.205}{.438} = .468$$

$$\beta_{YZ.X} = \frac{r_{YZ} - r_{XZ}r_{XY}}{1 - r_{XY}^2}$$

$$= \frac{(.918) - (.894)(.750)}{1 - (.750)^2}$$

$$= \frac{.248}{.438} = .566$$

b coefficients:

$$b_{XZ.Y} = \left(\frac{s_Z}{s_X}\right)\beta_{XZ.Y}$$

$$= \left(\frac{11.8}{2.2}\right)(.468)$$

$$= 2.510$$

$$b_{YZ.X} = \left(\frac{s_Z}{s_Y}\right)\beta_{YZ.X}$$

$$= \left(\frac{11.8}{88.0}\right)(.566)$$

$$= .076$$

a coefficient:

$$a = M_Z - b_{XZ.Y}M_X - b_{YZ.X}M_Y$$
$$= (18) - (2.510)(5) - (.076)(205)$$
$$= (18) - (12.55) - (15.58)$$
$$= -10.13$$

Multiple regression equation:

$$Z' = a + b_{XZ.Y}X + b_{YZ.X}Y$$
$$Z' = (-10.13) + 2.51X + .076Y$$

2. This prediction equation is statistically significant far beyond a chance level ($F = 750.0$, $d.f. = 2/97$, $p < .001$).
3. With a 32 percent margin of error, we know that the predicted viewing time is likely to fall within the range of $+2.9$ and -2.9 hours.
4. From further calculations, we can determine that approximately 42 percent of the prediction of viewing time is accountable to programming ratings and approximately 52 percent of this prediction is due to price of set.

Thus, for example, we could say that if the average programming rating (Variable X) was 4.0 for an individual and the price of set (Variable Y) was $180, then the estimated viewing time per week is 13.59 hours. Within 32 percent chance of error, we would estimate that this could vary between 16.49 and 10.69 hours. Or, for another example, we could say that with average programming ratings of 6.0 and a price of set equal to $330, the predicted viewing time is 27.73 hours, with a 32 percent margin of error of our estimate being between 30.63 and 24.83 hours.

THE LOGIC OF REGRESSION ANALYSIS

Equation for Multiple Regression

Similar to the regression equation presented in Chapter 12, the equation for multiple regression incorporates regression coefficients. Again, there is the a coefficient; but this time, we have two b coefficients, one for the X predictor variable and one for the Y variable. The equation has the general form:

$$Z' = a + b_{XZ.Y}X + b_{YZ.X}Y$$

In this case, the b coefficients are different for the two predictor variables. The $b_{XZ.Y}$ coefficient is the ratio of how many units Z increases for every unit increase in X, but with the effects of Y held constant. By contrast, $B_{YZ.X}$ is the ratio of how many units Z increases for every unit increase in Y, but where the effects of X are held constant. Because each of these two coefficients only reflects a portion of the prediction of Z, they are called *partial regression coefficients*.

The essence of the calculation is to find out what the a and the two b coefficients are equal to. As shown in Table 13.1, it is necessary to calculate first the *beta* (β) *coefficients*, or *standard partial regression coefficients*, prior to calculating the two b coefficients. These calculations are based on correlations found among the variables. The formulas for calculating coefficients are given in Table 13.1. Given the two β coefficients, the two b coefficients are then calculated. These calculations are also shown in Table 13.1. Calculation of the a coefficient incorporates the means of the three variables as well as the two b coefficients. This calculation is also shown in Table 13.1 for the sample data.

Interpretation of the Multiple Regression Equation

Given the solutions for the a and the two b coefficients for the sample data, we have the following prediction equation:

$$Z' = (-10.13) + 2.51X + .076Y$$
$$\text{(or)}$$
$$\text{Viewing time} = (-10.13) + (2.51 \times \text{programming rating}) + (.076 \times \text{price of set})$$

The foregoing should hold, for example, for what we know about the means of the three variables; that is, if $X = 5$ and $Y = 205$, then Z' should equal 18. That is,

$$Z' = (-10.13) + (2.51)(5) + (0.76)(205)$$
$$= (-10.13) + (12.55) + (15.58)$$
$$Z' = 18$$

Overall *F* Tests in Multiple Regression

The overall equation in multiple regression, as in the sample problem, can be tested for statistical significance. This is done by use of an F test, the same as discussed in Chapter 12 for a single-variable prediction equation. It is also essentially the same as using an F ratio to test the statistical significance of the multiple correlation coefficient associated with the multiple regression equation.

Note in Table 13.2 how the formulas for the calculation of F for either the regression equation or the multiple correlation coefficient are what was described in Chapter 12 as the ratio of "accounted for" (or "regression") variance to "unaccounted for" (or "residual") variance. They result in essentially the same F (within rounding errors), which in the sample problem is statistically significant $(p < .001)$. The practical interpretation is that we have a prediction that with very little probability could have arisen from chance alone. Or, in other words, the test provides evidence to reject the null hypothesis $(R = 0)$.

Standard Error of the Estimate

Just as in the simple linear regression example, the standard error of the estimate (SEE) reflects the standard deviation of actual values of the predicted variable (here, Z') for actual ones (Z). Calculation of SEE allows us to make a probability statement of the likelihood that an estimate will fall into some range of values about the predicted one. The SEE reflects the following relationship:

Table 13.2   Example Solution of an *F* Ratio for the Multiple Regression Equation or Multiple Correlation in the Sample Study

For the equation:

$$F = \frac{\text{(sum of squares, regression)} \div \text{(number of independent variables minus one)}}{\text{(sum of squares, residual)} \div \text{(sample size minus number of independent variables minus one)}}$$

$$F = \frac{12{,}947.3397 \div 2}{837.42022 \div 97} = \frac{6473.6695}{8.633198} = 749.9$$

For the multiple correlation coefficient:

$$F = \frac{\text{(multiple correlation coefficient squared)} \div \text{(number of independent variables minus one)}}{\text{(one minus multiple correlation coefficient squared)} \div \text{(sample size minus number of independent variables minus one)}}$$

$$F = \frac{(.9691)^2 \div 2}{(1 - .9691^2) \div 97} = \frac{.4696}{.0006268} = 749.2$$

$$SEE = \sqrt{\frac{\text{sum of squares residual}}{N - k - 1}}$$

where:

$$N = \text{sample size}$$
$$k = \text{number of independent variables}$$

Or, in the sample study:

$$SEE = \sqrt{\frac{837.42022}{97}} = 2.93 \text{ or } 2.9 \text{ (rounded)}$$

Contributions of Predictor Variables

We can also use the multiple regression equation to tell us how much each of the predictor variables contributes to the joint prediction of Z. Earlier (Chapter 11), it was noted that a squared multiple correlation coefficient tells us what proportion of the variance of the predicted variable is accounted for by the two predictor variables. In order to see the relative contribution of each predictor variable to this variance, we multiply each correlation coefficient of a predictor variable with the predicted variable (that is, r_{XZ}, r_{YZ}) by its respective beta (β) coefficient—for instance:

$$R^2_{Z.XY} = (\beta_{XZ.Y}r_{XZ}) + (\beta_{YZ.X}r_{YZ})$$

Using the sample data gives us:

$$(.9691)^2 = (.468)(.894) + (.566)(.918)$$
$$(.94) = (.42) + (.52)$$

Thus, we can conclude that programming ratings contribute .42 to the prediction of viewing time, and price of the television set contributes .52 to this joint prediction.

FURTHER NOTES ON PREDICTOR VARIABLES

In communication and education research, the achievement of a high degree of prediction (that is, a large R^2) is not always an end in itself. We could be concerned with the selection or elimination of predictor variables that might be useful or not useful as "causal" descriptions. We could be interested in how the relationships between predictors and predicted variables may vary by the inclusion of other variables. Or we might have the purpose of testing one alternative set of predictors against another. In brief, it is helpful to be aware of some of the procedures by which predictor variables are evaluated. We have already seen one example in the last section. This was in how the total prediction variance (R^2) could be broken down into components of variance (product of predictor variable correlations times β coefficients) contributed by predictor variables. Following are some further considerations of predictor variables.

Evaluation of *b* Coefficients

It may be recalled from Chapter 12 that a *b* coefficient, in essence, defines the amount that a predicted variable should change for increments of a predictor variable (remember the *slope* in Fig. 12.1?). In multiple regression, *b* coefficients are called *partial* because they reflect only the relation of the predictor with the predicted variable with other variables held constant. These coefficients are, thus, one way of evaluating a predictor variable.

Although we shall not go into detail on the calculations, it is possible to calculate the standard error of a *b* coefficient. This is done in situations where the researcher is interested in the reliability of a predictor variable in a multiple regression equation. The coefficient will vary among samples, and calculation of its standard error provides an estimate of the amount of such variation.

By the same token, it is possible to test the significance of a *b* coefficient against the null hypothesis that in the population it is zero. This provides a statistical basis for deciding whether a variable, relative to other predictors, has a significant contribution to the equation.

Other F Tests

There are several ways to test the significance of the amount of regression variance accounted for by a predictor or even a subset of predictors in a multiple regression equation. They all have the same logic of comparing "accounted for" ("regression") variance with "unaccounted for" ("residual") variance. Such tests might include the significance of adding or excluding one or several variables from an equation, the significance of one or a set of predictors compared with others, or changes in a predictor or subset of predictors given changes in other variables in the equation. All such tests are easy to conceptualize, and today, many computer programs for multiple regression supply the basis for calculating them.

"Stepwise" Analyses

Using a number of recent computer programs for calculating multiple regression equations, it is possible to develop the equation one variable at a time until some criterion is reached which indicates that further predictors are unnecessary. This is called a "stepwise" procedure. Usually, the variables are selected in the order of their ability to contribute to the overall prediction. The criterion of adding an additional variable is often determined by whether its contribution to predicted variance is statistically significant (say, by F test). An important practical note in stepwise analyses is that if additional variables are included in the equation, they may change the relative contribution of variables added earlier. Some computational programs include the elimination of variables added earlier when a variable added later makes their contribution insignificant.

Emphasis on Predictor Variables Rather Than on R^2

One of the basic interpretations of regression equations is the degree of prediction of the dependent variable as characterized by a large R^2 and a low SEE. Another basic focus, and one coming into increasing use in communication and education research, is on an interpretable and practically useful configuration of predictor variables, perhaps even at the expense of maximizing R^2. Suppose, for example, that prediction of children's success in use of a "reading instruction" kit is based on IQ, sex, reading readiness score, teacher rating, time spent with the kit, and parental attitudes. Some of these predictor variables are more easily obtained (for example, sex) than others (parental attitudes); hence, any multiple regression equation that emphasizes the former predictors, even if it yields a somewhat smaller R^2, will be more useful than one involving "costly" variables. Also, the researcher or administrator may wish to emphasize predictor variables that are themselves "manipulatable" in the classroom, such as time spent with the kit. As another example, we could probably increase the predictability of a student's success in graduate studies if we added additional predictors based on in-depth interviews. However, such interviews are not always feasible and are generally costly, so we settle for *Graduate Record Examination* scores (or another test), undergraduate grade point average, and letters of recommendation.

USE OF BINARY OR DUMMY VARIABLES

In both Chapter 11 (pp. 142–143) and Chapter 12 (pp. 155–156), we discussed how categorical variables could be coded with "0" or "1" and entered into correlational or regression analyses, a procedure using binary or dummy variables. This procedure can be extended as well to multiple regression analysis. We shall discuss two applications of such variables of practical utility to the reader: (1) coding categorical questionnaire or survey data, and (2) coding independent variables to illustrate the similarity of multiple regression analysis to analysis of variance.

Categorical Variable Coding and Analyses

One of the statistical limitations of many questionnaire-survey studies is that data often include categorical or nominal scales (Chapter 3) that do not lend themselves directly to more powerful or complex statistical analyses. Examples of categorical variables often found in surveys are sex, ethnicity, living area, occupation, and type (rather than amount) of training. In many studies involving such variables, analyses are limited to simple frequency tables of "cross-tabulations." Suppose, for example, that we were interested in studying the relationship between categorical variables and level of newspaper readership as gauged in average number of hours spent reading a newspaper each day. Results could be presented in a variety of cross-tabulations, such as:

		Sex	
		Male	Female
Occupation	Labor	1.2 hours	1.1 hours
	Management	2.1 hours	1.8 hours

Although such tabulations may be useful for simple descriptive purposes, they become complicated and cumbersome as more variables are added, especially continuous variables (such as age, income, and amount of education).

If the dependent variable is continuous (as is "hours" in the example), the relation of it with other variables, categorical or continuous, can be assessed in a multiple regression equation. This involves coding categorical variables, as described in Chapters 11 and 12, usually with "1" if a category applies, "0" if not. For example,

Variables			
Male	Female	Labor	Management
0	1	1	0

The above represents coding for a female laborer. Given such coding as predictor variables, it would then be possible (1) to assess the overall predictability of newspaper reading from sex and occupation data, (2) to test the statistical significance of the relationship, (3) to calculate the SEE of the prediction, and (4) to estimate the relative contribution to the prediction of the two predictor variables. Moreover, it is not difficult to envisage adding further predictor variables, categorical or continuous.

Binary or dummy variables greatly enhance our ability to reason from categorical data.

VARIATIONS OF MULTIPLE REGRESSION ANALYSIS

Just as we have discussed the relationship between regression analysis and analysis of variance in Chapter 12, there are similar relations with several other statistical approaches, most of which come under the heading of *multivariate* statistics. For example, if categorical variables serve as intended dependent variables, a related statistical method is *multiple discriminant analysis* (Chapter 16). There is the extension of multiple regression analysis to accommodate multiple dependent variables, a technique called *multivariate multiple regression analysis* (see Kerlinger and Pedhazur, 1973, or Monge, 1978, in the Supplementary Readings at the end of this chapter).

TIME-SERIES ANALYSIS

Although not strictly a subtopic of multiple regression, *time-series analysis* often involves the use of related statistical methods in the assessment of effects or relationships observed among multiple observations taken across time. Suppose, for example, that you wished to study fluctuations in the number of subscribers to a major monthly magazine. The simple comparison of subscribers at two points in time might prove inaccurate. Many factors could be causing effects—for example, seasonal patterns of buying, unexpected fluctuations in the economy, or perhaps some unusual sales had been stimulated by special editorial matter. In all, we would gain our best analysis by seeing the sales campaign relative to the pattern of sales across time. Once we had controlled for known patterns of variation, we might then be able to determine whether there was a steady pattern of growth (or decline).

There are also applications where a research hypothesis applies to effects across time rather than the typical "one-shot" effects in an experiment. For example, if a new management strategy is introduced into a factory, the hypothesis might be one of accumulative increase of productivity, not a simple "jump." Time-series analysis provides a framework for such experiments so that the patterns before and after the introduction of the change variable (the "interruption") can be compared for differences.

Interest in time-series analysis has been growing in communication research. You can read more about it in the article by Peter Monge included in the Supplementary Readings to this chapter. Also, the *SPSS-X User's Guide* contains an entire chapter on the Box-Jenkins method of time-series analysis.

SUMMARY

Multiple regression analysis provides for the prediction of values of a variable, given values of two or more variables. The equation describes the arithmetic combinations of the predictors for the best estimate of the predicted variable. The equation can be tested for statistical significance by use of an F test. The standard error (*standard error of the estimate* or SEE) of the predicted values can be calculated. The equation for multiple linear regression is an extension of the two-variable simple regression equation. Much of the focus in multiple regression analysis is centered on predictor variables—their statistical significance in regression equations, effects of adding or deleting variables, and their practical interpretation. Binary or dummy coding allows categorical variables to be introduced as predictors in multiple regression equations.

SUPPLEMENTARY READINGS

Box, G. E. P., and C. M. Jenkins, *Time Series Analysis: Forecasting and Control.* Revised ed. San Francisco: Holden-Day, 1976.

Cohen, Jacob, and Patricia Cohen, *Applied Regression/Correlation Analyses for the Behavioral Sciences.* 2nd ed. Hillsdale, N.J.: Lawrence Erlbaum Associates, 1983. Chapter 1 is an excellent introduction to regression and correlation as data analysis strategies. Chapter 3 concentrates on multiple regression.

Davis, Dennis K., and Jae-Won Lee, "Time-Series Analysis Models for Communication Research," Chapter 14 in Peter Monge and J. Cappella (eds.), *Multivariate Techniques in Communication Research.* New York: Academic Press, 1978,

Ferguson, G. A. and Y. Takane, *Statistical Analysis in Psychology and Education.* 6th ed. New York: McGraw-Hill, 1989.

Howell, David C., *Fundamental Statistics for the Behavioral Sciences.* Boston: PWS-Kent, 1989.

Kerlinger, Fred N., and E. J. Pedhazur, *Multiple Regression in Behavioral Research.* New York: Holt, Rinehart and Winston, 1973. A must for any person working seriously with multiple regression and multivariate statistics.

Monge, Peter R., "Multivariate Multiple Regression," Chapter 2 in Peter Monge and J. Cappella (eds.), *Multivariate Techniques in Communication Research.* New York: Academic Press, 1978.

SAS Institute, Inc., *SAS User's Guide: Statistics.* 1982 ed. Cary, N.C.: SAS Institute, 1982. See the section on "Regression" (pp. 3–110).

Spence, Janet T., John W. Cotton, Benton J. Underwood, and Carl P. Duncan, *Elementary Statistics.* 4th ed. Englewood Cliffs, N.J.: Prentice Hall, 1983. See Chapter 6.

SPSS, Inc., *SPSS-X User's Guide.* 3rd ed. Chicago: SPSS, 1988. Chapter 45 describes a program for regression analysis.

PART SIX

ANALYSES OF COMPLEX DIFFERENCES AND RELATIONSHIPS

14

Cluster Analysis

The statistical methods in this final group of chapters allow for data analysis where the research involves multiple variables. In the present chapter, Cluster Analysis, we see how cases can be grouped into clusters based on their similarity on multiple variables. Chapter 15, Factor Analysis, describes how multiple variables might be reduced to fewer underlying variables. Chapter 16, Discriminant Analysis, describes how we can distinguish among groups, based on how a linear combination of multiple measures may discriminate among them. Because the methods in these chapters are computationally complex and require use of computer programs for practical calculations, we will concentrate on their logic and application rather than calculation. References for the latter are provided in each chapter.

Although the basis of cluster analysis is mathematically complex, its use is relatively simple. As in a subsequent example, suppose we have different measures of demographic characteristics and media use variables on a group of people. Can we use these measures collectively to cluster people into groups based on their similarities?

WHAT IS CLUSTER ANALYSIS?

Cluster analysis is a method for grouping ("agglomerating") cases based on their similarities on multiple measures. This is a kind of a "birds of a feather flock together" analysis where an array or group of cases is subdivided successively into more and more subgroups. You might, for example, wish to see how U.S. radio stations can be grouped in terms of their similarities on such variables as power, size of staff, advertising revenue, estimated worth, and the like. Cluster analysis calculates solutions that successively divide the array of cases into smaller and smaller groups, each new group presumably being more internally homogeneous than the last. The clusters are then examined, if one wishes, to determine the similarity of cases to one another within each cluster and how the cases collectively contrast with cases of other clusters.

A SAMPLE STUDY

As the public's media choices have proliferated, it has become interesting to see if there are particular "classes" of individuals according to media uses. Cluster analysis is useful for this type of study, not only because it gives a mathematical basis for grouping cases on multiple variables but also because it provides a hierarchy of clusterings that allow you to determine how much differentiation contributes best to the purposes of your study. That is, how finely do you want the "classes" divided?

Problem

Our problem involves analysis of demographic and media use data from 101 individuals hypothetically sampled randomly from a small community of persons 18 years of age or older. Demographic variables include age, sex, education, and income. Media use variables are number of hours weekly of magazine reading, book reading, newspaper reading, watching broadcast television, watching VCR programs, and whether or not an individual uses a personal computer.

The research question is: Can individuals from . . . be divided into meaningful subgroups in terms of a combination of demographic and media use variables?

Method

A summary of the descriptive statistics from the data set is given in Table 14.1. Note from this summary that the sample reflects an older group (the community includes retirees; the sample had 10 people of 60 years or older). They tend to be well educated; only nine were not high school graduates. Their income is above average, the state average being about $18,000 at the time. They watch an average of 4.3 hours of television daily, about 1 hour less on the average than annual figures for their age group at the time. Their use of a VCR is considerably less; most watch a movie only occasionally. They spend a little less than 1 hour on the average each day reading newspapers. Most are book and magazine readers. About half of the individuals use personal computers.

Table 14.2 provides information on intercorrelations among the above variables. Some generalizations based on Table 14.2 include the following. Similar to other studies of these interrelations, use of print media was positively correlated with education and was negatively correlated with hours of television watching (but not VCR use). Income showed a partly

Table 14.1 **Variables and Descriptive Statistics of the Media Use Study ($N = 101$; Simulated Data)**

Variable	Mean	Standard Deviation
Age (years)	44.6	15.9
Sex (female = 0; male = 1)	.495	.50
Education (years)	14.4	3.1
Income ($ yearly)	30,079	27.3
Magazine use (hours weekly)	3.9	2.2
Book use (hours weekly)	4.5	2.6
Newspaper use (hours weekly)	6.2	3.6
Broadcast television (hours weekly)	30.3	22.1
VCR television (hours weekly)	3.4	4.8
PC use (0 = no; 1 = yes)	.51	.50

Table 14.2   Intercorrelations among Demographic and Media Use Variables

	Sex	Educ.	Inc.	Mag.	Book	News	TV	VCR	PC
Age	.03	−.21	−.09	.03	.02	−.04	.19	−.21	−.57
Sex		.02	.13	−.07	−.07	−.04	.02	−.19	−.05
Education			.35	.54	.32	.21	−.50	.24	.22
Income				.11	.00	.35	−.20	.09	−.07
Magazine					.57	−.02	−.15	.06	.08
Book						.06	−.16	−.08	.13
Newspaper							−.03	.00	−.07
TV								−.11	−.01
VCR									−.01

similar pattern relative to newspaper and television use. Personal computer use varied inversely with age, older individuals not being users. More education, by contrast, was slightly related to use of personal computers. The greatest correlation among media use variables was the r of .57 between hours spent with books and hours with magazines, a relation often observed in studies of media uses. However, unlike some studies, the inverse correlation between print and broadcast use was present but small (books and magazines with TV, −.16 and −.15, respectively).

Based on these observations, we would expect to have some bases for clustering individuals in terms of the mix of demographics and media uses. It was not necessary to compute a correlation matrix to do a cluster analysis; we did this only to give a general "picture" of the data. The cluster analysis was performed (*SPSS-X User's Guide*, 1988, Chapter 23), and the results are reviewed below.

Results

As already mentioned, cluster analysis goes through a series of stages ("agglomeration schedule") successively reclustering the cases into typically smaller and more homogeneous groups. Successive groupings are based upon "proximity measures" that the researcher selects for calculation. Cases are clustered in terms of such proximities at each stage; then, a next set of proximities is calculated, and another stage is defined. The analysis starts with all cases combined, and continues through as many stages as the researcher selects, the last possible one of which would equal the number of cases.

The easiest way to view these successive stages is in terms of an "icicle" plot. Figure 14.1 illustrates part of a vertical icicle plot for the present analysis. (It is also possible to display the successive stages horizontally.) In interpreting these results, it is necessary to refer to the characteristics of the cases.

Note that in the first division, in Stage 2, of the analysis (Fig. 14.1), the clustering is to group Cases (individuals) 55, 12, and 11, as against all

FIGURE 14.1 A portion of the vertical icicle plot for media use cluster analysis

```
Cluster                              Cases
Stage                   (19 of the 101 individuals by ID number)*

                    1
       5  1  1  0  9  8  4  5  1  6  5  1  1  6  5  3  6  9  4
       5  2  1  1  8  7  0  6  5  4  3  4  3  5  9  8  3  5  7

  (1) xxxxxxxxxxxxxxxxxxxxxxxxxxxxxxxxxxxxxxxxxxxxxxxxxxxxxx
  (2) xxxxxxx  xxxxxxxxxxxxxxxxxxxxxxxxxxxxxxxxxxxxxxxxxxxxx
  (3) xxxxxxx  xxxxxxxxxx  xxxxxxxxxxxxxxxxxxxxxxxxxxxxxxxxx
  (4) xxxxxxx  xxxxxxxxxx  xxxxxxxxxxxxxxx  xxxxxxxxxxxxxxx
  (5) xxxx  x  xxxxxxxxxx  xxxxxxxxxxxxxxx  xxxxxxxxxxxxxxx
  (6) x  x  x  xxxxxxxxxx  xxxxxxxxxxxxxxx  xxxxxxxxxxxxxxx
  (7) x  x  x  x  xxxxxxx  xxxxxxxxxxxxxxx  xxxxxxxxxxxxxxx
  (8) x  x  x  x  xxxxxxx  xxxx  xxxxxxxxx  xxxxxxxxxxxxxxx
```

* The 81 cases not shown (they would be to the right of this plot figure) were as yet undifferentiated in the analysis; that is, an "x" would appear in the plot below all other cases in the first eight stages.

remaining cases (not all shown here). Examination of the data for these individuals shows them to be nonwatchers of television (0 hours), relatively high income (two were $90,000; the third was $89,000), and high levels of education (two were 18 years; the third was 16). It appears, then, that the first cluster to be identified in the analysis was mainly three individuals (two men, one woman) who were very similar to one another but who also contrasted considerably with the remainder of the subjects.

Stage 3 includes Subjects 101, 98, 87, and 40 in a new cluster. They are all females, with considerably higher than average television viewing (40 hours), no use of personal computers or VCRs, and lower-than-average education (10 years).

Stage 4 clusters Subjects 56, 15, 64, 53, and 14, all of whom are similar in having a high income average ($84,000).

Note in Stage 5 that the only change is to divide Subject 11 from the earlier Stage 2 cluster. This is the female subject of the three; she watches VCR programming 16 hours per week, whereas the other two watch only 1 and 0 hours each. In Stage 6, the other two members of this group are divided from one another.

In Stage 7, Subject 101 is divided from other clusters, presumably because her education level was higher than some and lower than other clusters. Stage 8 subdivides the group clustered in Stage 4, apparently in terms of print media uses.

The results of the cluster analysis tell us that rather than reflecting several large cluster types, the present sample comprises more in the way of a few small homogeneous clusters and a large undifferentiated one. Beyond Stage 8, most new clusters are two or three cases each, and after Stage 35, most new clusters are individual cases.

SOME PRACTICAL POINTS

It is important to bear in mind that cluster analysis literally measures the similarity between cases and makes the cluster distinctions at points of least similarity in the stages of the analysis. Like many analyses involving multiple variables, it cannot interpret any social or psychological "sense" from the variables, nor is there any guarantee that the clusters will be useful. If children had been included in the present sample, chances are that they would have been differentiated early as a major cluster because they are very similar to one another in demographics and media uses, especially in comparison with adults.

As resulted in the present data set, most early clusters were small and reflected people who were similar in both demographics and media uses, often in combinations usually found in studies of this type (for example, inverse correlation of education and income with television viewing or positive correlation with use of print). Also, these early clusters tended to be distinguished relative to the general statistics of the sample—for example, means in television viewing, income, or education. Practically speaking, in terms of these types of variables, many of the people in the sample community tended to fall into one common group with many small variations; also, some were clustered into small groups of quite similar people who tended to vary substantially from the main group.

It would be instructive to become familiar with other ways in which these multiple demographic and media use variables can be used to make generalizations about this population. Examples of factor and discriminant analyses of this same data set are included in the next two chapters.

SUMMARY

Cluster analysis is a statistical "agglomeration" procedure whereby cases can be identified as belonging to clusters based upon comparisons of similarity measures calculated among cases. The analysis proceeds in a sequential manner whereby successive stages subdivide the overall array into smaller and more homogeneous subclusters. Cluster definition is strictly by mathematical criteria. Determining which stage of clusters is meaningful is left to the subjective interpretation of the researcher.

SUPPLEMENTARY READINGS

Anderberg, M. J., *Cluster Analysis for Applications*. New York: Academic Press, 1973
SPSS, Inc., *SPSS-X User's Guide*. 3rd ed. Chicago: SPSS, 1988. Chapter 23 describes a computer program for cluster analysis.

15

Factor Analysis

If a researcher has a complex array of measures, it is sometimes useful to determine whether they can be combined into fewer, more basic, or underlying variables. For example, one might find that measures of vocabulary recognition, reading speed, ability to use grammatical rules, and content recall are highly intercorrelated and readily combinable into one overall variable of "verbal ability." In this chapter, we discuss the basic use of factor analysis for this purpose, including some of the logic underlying the strategies involved. In addition to a sample problem introduced in this chapter, the demographic and media use variables described in Chapter 14 are factor analyzed to illustrate the contrast with cluster analysis.

A PRACTICAL VIEW OF FACTOR ANALYSIS

In the three previous chapters, we have seen how relationships among measures are assessed in terms of the variance that they have in common. We saw in Chapter 11 how common variance was the underlying concept in describing correlation—an index of the degree and direction to which measures vary together. In Chapters 12 and 13, we saw how a knowledge of common variance between or among measures could be used in formulas for the prediction of measurement values of one variable, given a value or values of other related variables. In both correlation and regression, we were concerned with what might generally be called *measurement overlap*.

In addition to assessing the correlation between measures, or predicting one measure from another, we can view measurement overlap as the extent to which measures of different variables are measuring something in common. Factor analysis is a mathematical procedure applied for this purpose. It aids us in answering the question: Given a relatively large number of variables, does their measurement overlap, indicating that there may be fewer, more basic, and unique variables underlying this larger number? Factor analysis procedure takes the variance defined by the intercorrelations among a set of measures and attempts to allocate it in terms of fewer underlying hypothetical variables. These hypothetical variables are called *factors*.

A SAMPLE STUDY

Problem

Suppose, for example, that we have obtained a number of different measurements all presumed relevant in one way or another to the use of an educational television series and classroom activities both aimed at reducing sex-role stereotyping on the part of elementary school students. Our aim is to determine whether there is an interpretable underlying structure which can be found in the relationships among these variables.

Method

For purposes of illustration, let us simulate a study with the following variables:

1. *Student's IQ* as obtained from a standardized test
2. *Male stereotype score*, a measure of the degree of male stereotyping a student imposes when estimating whether males, females, or both would fill a given occupation
3. *Female stereotype score*, the degree of female stereotyping in estimating who would fill an occupation
4. *Recall of TV program content*, scores on a multiple-choice test about the TV programs
5. *Comprehension of instructional objectives*, measuring the student's understanding of the problems of imposing stereotypes as shown in the TV programs and activities
6. *Reasoning from instructional objectives*, the ability to draw inferences about the negative effects of stereotypes
7. *Student's liking of materials*, scaled ratings of the TV programs and activities
8. *TV-viewing frequency*, number of programs viewed
9. *Classroom activity frequency*, number of exercises in which a student has participated
10. *Teacher's attitude toward the materials*, scaled ratings by the child's teacher of favorability of attitude toward the TV program and activities; this rating is assigned to each child's set of measures

We would expect that a number of the foregoing measures would be interrelated ("intercorrelated"). For example, the different measures involving the understanding and reasoning from the program content would probably be interrelated, perhaps reflecting some overall comprehension aspect. Similarly, we might expect that variables such as television-viewing frequency or classroom activity frequency would be related to some of the outcomes of the study, such as the attitude or comprehension measures. Recall how in Chapter 11 that interrelationships were graphically portrayed in terms of overlapping circles showing common variance (Figs. 11.3 and 11.4). Using this same system of circles, we might picture the overlap among the foregoing 10 variables as shown in Figure 15.1.

The overlap among circles (Fig. 15.1) can be thought of as the proportions of variance that the various measurements have in common, or the square root of each of these proportions of common variance could be depicted as correlation coefficients as summarized in Table 15.1. We can look at

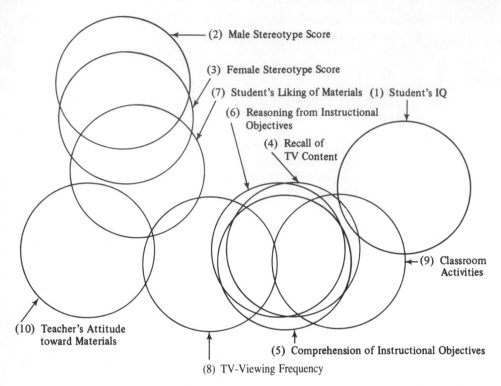

FIGURE 15.1   Measurement overlap of all correlations greater than .45 (with some distortions to facilitate illustration)

the circles in Figure 15.1 or the correlation coefficients in Table 15.1 measure by measure if we wish, and we can see what each measure has in common with each other measure. From a broader viewpoint, however, we might consider how all 10 measures tend to cluster about one another. We can think of each cluster, or any circle not in a cluster, as perhaps reflecting some aspect of the behavior that we are assessing. That is, our measures of recall, comprehension, and reasoning do, indeed, seem to be measuring roughly the same thing—something that we might more generally call basic comprehension. We can note also how the two different measures of stereotype bias in occupations, one for males and one for females, might simply be representing an overall bias toward stereotyping. By the same token, we can see in Figure 15.1 or in Table 15.1 that the measure of student's IQ does not seem highly interrelated with any other of the variables, save for a moderate-to-low correlation with classroom activity frequency. Finally, it is evident that some of the "causal" variables are interrelated with "effect" variables, such as television-viewing frequency with the comprehension measures.

These clusters of measurement overlap do tend to show us more of a

Table 15.1 **Intercorrelation Matrix for Variables in Use of Television to Reduce Sex-Role Stereotyping (Simulated Study)**

Variables	Correlation with Each Other (Variables Designated by Number)								
	(2)	(3)	(4)	(5)	(6)	(7)	(8)	(9)	(10)
(1) Student's IQ	.07	.07	.31	.16	.20	−.07	−.01	.45	−.04
(2) Male stereotype score		−.89	.06	.34	.08	−.37	.37	−.15	.30
(3) Female stereotype score			−.11	.43	.09	−.57	.30	−.07	.21
(4) Recall of TV content				.89	.97	.10	.68	.83	−.03
(5 Comprehension of instructional objectives					.83	−.10	.67	.70	.00
(6) Reasoning from instructional objectives						.11	.67	.70	.00
(7) Student's liking of materials							.23	−.08	.46
(8) TV-viewing frequency								.19	.19
(9) Classroom activity frequency									−.22
(10) Teacher's attitude toward materials									

basic, underlying structure than do the 10 individual variables. We could, thus, reduce our description of this study from 10 variables to fewer variables and have a more basic and more economical way to characterize the behavior we are studying.

What we have done in plotting the relations among the various measures as in Figure 15.1 is essentially what the mathematical technique of factor analysis provides. Put into practical terms, factor analysis takes a matrix of intercorrelations such as shown in Table 15.1 and provides a method for determining, on the basis of the correlations among a relatively large

Table 15.2   Factor Matrix (Varimax Rotation) for the Intercorrelations Shown in Table 15.1 and Figure 15.1

	Factors		
Variables	I	II	III
(1) Student's IQ	.31	.12	−.29
(2) Male stereotype score	−.06	−.90	−.21
(3) Female stereotype score	−.09	−.97	−.03
(4) Recall of TV content	.99	−.01	.01
(5) Comprehension of objectives	.88	−.34	.01
(6) Reasoning from objectives	.97	−.02	.07
(7) Student's liking of materials	.10	.60	.69
(8) TV-viewing frequency	.68	−.27	.52
(9) Classroom activity frequency	.84	.16	−.42
(10) Teacher's attitude toward materials	−.04	−.13	.66

number of variables, what fewer, more basic, and unique variables might underlie this larger number of variables. A factor analysis solution was conducted for these intercorrelations, and the results are summarized in Table 15.2

Notice first that the factor matrix (Table 15.2) lists the factors as columns I, II, and III. This indicates, so to speak, that three dimensions of common variance were found in the analysis (similar to the three clusters portrayed in Fig. 15.1). The numbers given within the matrix itself are interpreted as correlation coefficients. They tell us the correlations between each of the original measurements and each of the three factors, or hypothetical underlying variables found in the analysis. Thus, we can see, as was the case in Figure 15.1, that the measures of recall, comprehension, and reasoning from program content all tend to be highly correlated with what is defined as Factor I, whereas the two attitudinal measures of stereotyping tend to be highly correlated with Factor II. When we consider a variable's correlation with a factor, we talk about this correlation in terms of *factor loading*. That is, we would say that the measures of recall, comprehension, and reasoning from objectives are all "highly loaded" on Factor I and that the stereotype measures "load" on Factor II.

Results and Interpretations

A *factor matrix* (or, as discussed later, a *rotated matrix*) is mainly what the mathematics of factor analysis provides us. The interpretation of what each factor means, or how we might label it, is left to the subjective evaluation of the researcher. Usually, this evaluation involves close examination of the factor matrix and asking such questions as: What variables tend to load highest on each factor? How might these variables together define a more basic variable relevant to what is being studied? In interpreting the factor matrix shown in Table 15.2, we might report the following conclusions:

Three factors were extracted in the analysis. These were labeled as follows: (I) *comprehension/exposure*, defined by dependent variable measures of recall, comprehension, and reasoning, and independent variables of exposure to the TV program and classroom activities; (II) *stereotyping/student's liking of materials*, defined by the inverse relationship of the degree to which students would impose sex-role stereotypes upon occupations and the degree to which they indicated liking of the program and activity materials; and (III) *teacher's attitude/student's liking of materials*, defined as the simple relationship between those two particular variables. For subsequent analyses, it appears useful to combine the dependent-type variables of recall, comprehension, and reasoning into a single underlying variable designated as "understanding." It also appears useful to combine the separate measure of male and female stereotyping into one overall "stereotyping" score.

Another point concerning the interpretation of the factor matrix is that factor loadings (as in Table 15.2), like correlation coefficients, have signs. Usually, unless a minus sign appears, the relationship between the variable and the factor is assumed to be positive. That is, the plus signs are simply omitted. Such signs, as was the case with correlation coefficients, are indicative of the direction of the relation. Thus, the more that the students indicated a liking of the materials, the less they tended to impose sex-role stereotypes in their assignments of people to occupations (see Factor II). The variables with high negative loadings are just as pertinent to factor interpretations as are the variables with high positive loadings. Loadings with different signs on the given factor mean that the variables' scores have different directions of relationship.

Probably the most frequent use of factor analysis in communication and education research is where the researcher wishes to determine whether a relatively large number of measures can be reduced to fewer, more basic, and underlying variables, as in how items on a test of some type may represent fewer variables than the items themselves, as was the case with the measures of recall, comprehension, and reasoning, or the stereotyping scores, in the foregoing example. If you have taken achievement examinations, you probably know that the test items on such exams reduce to factors such as mathematical or verbal reasoning ability. A frequent use of factor analysis is similar to the present example where there is a desire to see what type of structure underlies the variables in a study. Parts of a factor structure may simply represent redundancy among certain measures or relationships of causality in others. We have chosen this latter example because we shall use parts of it in subsequent chapters to show relationships among different multivariate approaches.

THE LOGIC OF FACTOR ANALYSIS

Although the detailed mathematical logic of factor analysis is too complex a topic to present here, we can provide a rather general view of what a factor is and, furthermore, of how we can rotate factor structures in order to provide the best fit, or most useful fit, of the factors to the data or to whatever theoretical structure is being considered.

Factors and Rotation

As was said earlier, factors are a redefinition of the interrelations among variables according to how these interrelations might be defined in terms of fewer (hypothetical) variables. In order to illustrate what is meant by this, let us consider a simplified problem involving five variables and eventually two factors.

Consider first the matrix of intercorrelations presented in Table 15.3 (A). The correlation coefficients define for us the interrelations among the variables A through E. Simply by inspection of this table, we intuit that Variables A and B tend to vary together more so than either with Variables C, D, and E and that C, D, and E seem to go together. We could plot these interrelations in terms of overlapping circles (as in Fig. 15.1) if we

Table 15.3   Simplified Example of (A) Intercorrelation Matrix, (B) Unrotated Factor Matrix, and (C) Rotated Factor Matrix

(A) *Intercorrelations*

		Variables			
		B	C	D	E
	A	.985	.500	.342	.174
Variables:	B		.642	.500	.342
	C			.985	.940
	D				.985

(B) *Unrotated Factor Matrix*

		Factors	
		I	II
	A	.574	.819
	B	.423	.906
Variables:	C	−.423	.906
	D	−.574	.819
	E	−.707	.707

(C) *Rotated Factor Matrix*

		Factors	
		I	II
	A	.996	.087
	B	.966	.259
Variables:	C	.423	.906
	D	.259	.966
	E	.087	.996

wanted to. For present purposes, however, a better way to plot these intercorrelations is shown in Figure 15.2 (A). Each arrow (vector) here represents one of the variables. The angle between any two variables reflects the correlation between them. The smaller the angle between

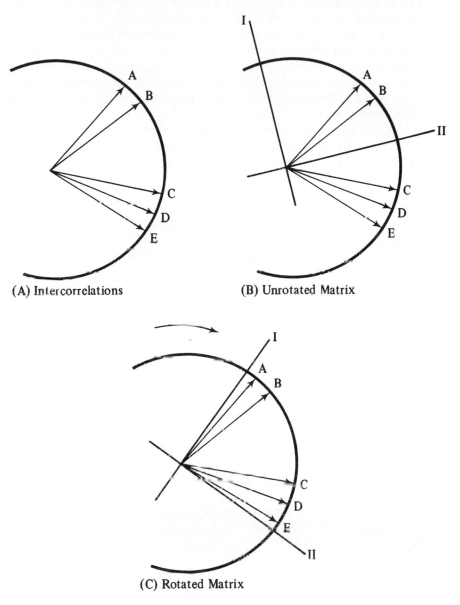

(A) Intercorrelations   (B) Unrotated Matrix

(C) Rotated Matrix

FIGURE 15.2   Graphic portrayal of (A) intercorrelations, (B) unrotated matrix, and (C) rotated matrix for data given in Table 15.3. The correlation between any two vectors is equal to the cosine of the angle between them. Variables are Vectors A through E; factors are Vectors I and II.

any two vectors, the greater is their correlation. (In order to provide this illustration, the cosine of each angle between two variables was taken to be equal to its respective correlation coefficient. For example, the angle between A and B is 10° and its cosine is .985, the correlation between A and B; or the angle between A and E is 80° and its cosine is .174, the correlation between A and E. This has nothing to do with the logic of factor analysis—it is simply necessary to provide this illustration.)

What we have, then, in Figure 15.2 (A) portrays only the interrelations among the five variables. By measuring the angles between all possible pairs of vectors, we could provide the correlation coefficients shown in Table 15.3 (A). These coefficients are what we begin with in conducting a factor analysis.

Suppose, next, that on the basis of certain mathematical criteria (which differ for various types of solutions), we considered the relations among the five variables not in terms of interrelations among each other but all in relation to two factors (I and II). For purposes of illustration, suppose that these two factors were initially defined as shown in Figure 15.2 (B). A first point to consider is that the correlations between the variables and the two factors are reflected in the angles (actually cosines of the angles) between each of the vectors representing variables and the two vectors now representing factors. We can still define the relations among the variables, but now, we do it in terms of the two factors as reference points. In other words, we can define the relations among the variables in terms of how they correlate with the two factors. These correlations, or factor loadings, are presented in Table 15.3 (B), called an *unrotated factor matrix*. Another consideration at this point is that the two factors are treated as completely uncorrelated with each other (that is, 90° apart in Fig. 15.2 [B], and the cosine of 90° equals 0, their correlation). This is a common assumption in many types of factor analysis solutions but is not a necessary assumption by any means.[1]

If we want these two factors to reflect as clearly as possible what we might consider to be a hypothetical variable underlying Variables A and B and another underlying Variables C, D, and E, we rotate their position (keeping them 90° apart) as shown in Figure 15.2 (C). Notice now that we have made the best compromise possible in locating Factor I with as small angles (or as large a correlation) as possible in locating it near Variables A and B while at the same time positioning Factor II as closely as possible to Variables C, D, and E. When we now consider the correlations of the five variables with the two factors, the pattern is relatively clear-cut. In Table 15.3 (C), notice that Variables A and B have relatively high loadings on Factor I (.996 and .966, respectively), whereas they have relatively lower loadings on Factor II (.087 and .259, respectively). Similarly, Var-

---

[1] Treatment of factors as independent of one another is called an orthogonal solution, whereas consideration of factors as interrelated is called an oblique solution.

iables C, D, and E all have relatively high loadings on Factor II (.906, .966, and .996, respectively) but have relatively low loadings on Factor I (.423, .259, and .087, respectively).

The rotation, then, has provided a readily interpretable factor structure, as can be seen by assessing the factor loadings in the rotated factor matrix in Table 15.3 (C). We would probably conclude that two more basic variables, as defined by Factors I and II, underlie the five measurements represented in Variables A through E. In an actual case of using factor analysis, we would use this rotated factor matrix as a basis for reporting our conclusions (as we did with reference to Table 15.3).

A word of warning is in order, however. What we have seen in the foregoing example is a highly simplified illustration of factors and of factor rotation. For one thing, the illustration said nothing about measurement error. That is, presumably some of the variance in a collection of measurements is due to error and, thus, must be somehow defined in the factor analysis solution. Neither did the illustration accommodate the fact that variables typically have some portion of their variance that is specific to them alone and may not be incorporated into the factor structure. Additionally, the preceding illustration gives no hint of the computational procedures involved in factor analysis. There are mathematical criteria (as well as some subjective decisions necessary to use these criteria) involved in determining the number of factors to be extracted, in locating these factors prior to rotation, and in rotating the factor structure. Even if we were to undertake a computational solution of the present simple problem, it would take considerable time for hand calculation. Evidence of the computational complexity of factor analysis solutions is that many of our current methods would not be economically feasible (or even possible in many cases) were it not for the development of electronic computers. Finally, factor analysis solutions often involve extraction of more than two factors. Obviously, we would be hard pressed to present graphically more than two factors at a time when using illustrations like Figure 15.2; the present illustration is a simple one indeed. At best, all that has been done is to present a rudimentary picture of the relations between a correlational matrix and unrotated and rotated factor matrixes.

Relation between the Correlation Matrix and the Factor Matrix

Another way of viewing the essence of factor analysis is to consider the relationship between what is defined in the correlation matrix and what is defined in the factor matrix. Recall that we have characterized factor analysis as the redefinition of the interrelations among variables in terms of relations of these variables with a set of (fewer) hypothetical variables. If we consider that both the correlation and factor matrixes are two different ways of defining the same interrelations among a set of variables, then we can define the statistical goal of factor analysis as the substitution of a factor matrix for a correlation matrix. We can get some idea of the relation between these matrixes by comparing parts (A), (B), and (C) in Table 15.3. (Again, bear in mind that the data here are fictitious and incorporate no measurement error.) We can state a relation as follows: The correlation between any two variables as shown in the correlation matrix (for instance, r_{AB}) is equal to the sum of the products of the two variables' loadings on each factor, as shown in any form, unrotated or rotated, of the factor matrix [for example, $(A_I B_I) + (A_{II} B_{II})$].[2] No doubt, this statement seems awkward and confusing at first. (If we took the time to define the necessary mathematical notations, we could state it much more parsimoniously.) What it simply says, for example, is this: If we take the product of the loadings of Variables A and B on Factor I, say, $.574 \cdot .423$ in the unrotated matrix in Figure 15.2 (B) and add it to the product of the loading of these same two variables on Factor II ($.819 \cdot .906$), then this sum will equal the correlation between Variables A and B ($r_{AB} = .985$ in the correlation matrix). The same holds true between our rotated matrix and the correlation matrix—that is,

$$(A_I B_I) + (A_{II} B_{II}) = r_{AB}$$
$$(.996 \cdot .966) + (.087 \cdot .259) = .985$$

For another example, consider the correlation between Variables C and E ($r_{CE} = .940$):

$$(C_I E_I) + (C_{II} E_{II}) = r_{CE}$$
Unrotated: $(-.423 \cdot -.707) + (.906 \cdot .707) = .940$
Rotated: $(.423 \cdot .087) + (.906 \cdot .996) = .940$

Another way, then, of viewing the logic of factor analysis is simply that of matrix substitution, or what we have called redefinition.

Rotational Schemes

In the present example, we rotated the two factors so that we could get maximal loading of Variables A and B on Factor I and getting maximal loading of Variables C, D, and E on Factor II. Or we could look at this in terms of placing A and B high on Factor I, but placing C, D, and E low on this factor, or placing C, D, and E high on Factor II, but A and B low on this factor. This kind of a rotation comes close to illustrating what we call rotation to *simple structure*. There are, however, a number of different procedures for factor rotation. Suppose, by contrast, that our theory led to the prediction that there should be two factors and that certain variables would load on each. In this case, we would rotate the

---

[2] Where A_I means the loading of Variable A on Factor I; B_{II} the loading of Variable B on Factor II; and so on.

axes to where theory suggests that they fit; this is called rotation to *psychological fit.*

One facet of research on factor analysis methods has centered on development of rotational schemes that are as objective as possible. Two of these schemes that attempt to objectify mathematically the criterion of simple structure are the *quartimax* and *varimax* methods.

Sources of Variation

Earlier, it was mentioned that our simplified example said nothing about measurement error or about how we accommodate nonerror variance that is not incorporated in a factor structure. We can think of both of these aspects as potential sources of variation to be treated in a factor analysis. For present purposes, let us label them as:

V_e = error variance

V_s = nonerror variance specific to variables and not included in the factor structure

In addition to the above sources of variation, we have that variance which is incorporated into the factors defined in the analysis; we call this *common factor variance,* or V_c. Common factor variance is, in turn, composed of the variance attributed to *each* of the factors determined in the analysis— that is,

$$V_c = V_I + V_{II} + V_{etc.}$$

We consider that all of the foregoing sources of variation together define what we call *total variance* or V_t. Thus, for a given analysis, we can conceive of a total amount of variance (V_t) being divided into the various sources of variation. In a three-factor structure, for example,

$$V_t = V_I + V_{II} + V_{III} + V_s + V_e$$

Table 15.4 presents a complete matrix for a hypothetical solution involving five variables (M through Q) and where three factors (I, II, III) have been extracted. With five variables, the total variance (V_t) would equal 5.0. By examining this matrix, we can see the division of this total variance into its various sources. In terms of actual values of variance, we can see how they are divided among the factors (V_I, V_{II}, V_{III}) as well as according to specific variance (V_s) and error variance (V_e), thus,

$$V_t = V_I + V_{II} + V_{III} + V_s + V_e$$
$$5.0 = 1.74 + 1.25 + 1.23 + 0.37 + 0.41$$

Table 15.4   Example of a Rotated Factor Matrix Including All Sources of Variation

		Factors			Sources of Variation		
		I	II	III	(V_c)	V_s	V_e
	M	.10	.10	.90	(.83)	.07	.10
	N	.08	.65	.65	(.85)	.08	.07
Variables:	O	.95	.06	.04	(.91)	.06	.03
	P	.90	.20	.06	(.85)	.01	.14
	Q	.08	.88	.02	(.78)	.15	.07
Variance:		1.74	1.25	1.23	(4.22)	0.37	0.41
Proportion of total variance:		.35	.25	.25	(.85)	.07	.08
Proportion of common variance:		.41	.30	.29			

It is usually more convenient, however, to view these sources of variation in terms of *proportion of total variance;* thus,

$$V_t = V_I + V_{II} + V_{III} + V_s + V_e$$
$$1.0 = .35 + .25 + .25 + .07 + .08$$

Given the above overall view, the interest then centers on what has been defined as common factor variance (V_c). In Table 15.4, V_c is equal to 4.22 in terms of actual variance, and this is approximately .85 of the total variance. In focusing on common factor variance, it is additionally useful to understand how the individual factors each account for this variance; this is called *proportion of common factor variance* (as contrasted with the proportion of total variance considered above for each factor). In terms of such proportions, we would say that:

$$V_c = V_I + V_{II} + V_{III}$$
$$1.0 = .41 + .30 + .29$$

What we have described thus far in Table 15.4 has centered only on the overall sources of variation. We can next examine the matrix in more detail and see how the individual variables (M through Q) reflect sources of variation. Recall that factor loadings are the correlations of each variable with a factor. Like correlation coefficients, these loadings when squared indicate the variance that the two variables have in common. For example, if the loading of Variable M on Factor I is .10, then we can say that this contributes .01 (that is, $.10 \cdot .10$) as a source of variation. Notice in Table 15.4 that if we take the three factor loadings for Variable M (.10, .10, .90) and square them (.01, .01, .81) we have the amount of common variance (.83), or *commonality,* attributed to that variable. In other words, on the basis of the factor loadings shown in the matrix, we can break down the sources of common variance according to the relations between the variables and the factors.

It might be worthwhile to digress for a moment and to consider how the foregoing breakdown would apply to the earlier sample of the rotated factor matrix shown in Table 15.3. Remember, we noted earlier that neither specific variance nor error was considered in this example. Indeed, what we actually assumed was that there was no error or specific variance. In this case, the sum of the squared factor loadings on each variable should equal 1.0, because the total variance in this matrix is all common variance. This, indeed, is the case: for example, taking the loadings of Variable A on Factors I and II in the rotated matrix (Table 15.3), we have .996 and .087. When these are squared (and rounded), we have .99 and .01, which when added are equal to 1.0.

Returning to Table 15.4, we can also see how the individual factor loadings can be seen in relation to the variance attributable to each factor extracted in the analysis. This time, we square the factor loadings and sum *down* the column. Using Factor I in Table 15.4 as an example, we take the loading of each variable (.10, .08, .95, .90, .08), square each (.01, .0064, .9025, .81, .0064), then sum, and we have the amount of variance attributable to this factor (1.74). In analyses such as this, the sum of squared factor loadings is called an *eigenvalue*.

What we have seen, then, in all of the foregoing is how a factor analysis solution can be interpreted in terms of the various sources of variation. We can consider the overall picture in terms of variance attributable to common variance (or the individual factors), to specific variance, and to error variance. Or we can consider variance attributable to each factor in terms of its relative proportion of common variance. In more detail, we can consider how the individual variables contribute as sources of variance. In this case, the contribution of each variable is taken as the square of its loading on each factor.

SOME PRACTICAL ASPECTS OF FACTOR ANALYSIS

Given a set of measurements on a number of variables, what are some of the practical considerations reflected when factor analysis is used? Because we have not gone into any detail in terms of computational procedures, we can do no better than to discuss some of the high points of these considerations.

Measurement Error

Given a matrix of intercorrelations among the variables to be factor analyzed, one problem concerns defining the error variance (V_e) attributable to each variable. Suppose that, on the basis of prior research, we knew how much nonerror variance was attributable to a given variable. We could describe this nonerror variance on the basis of determining the correlation of the variable with itself in some type of test, retest situation. The square of this correlation coefficient would then be what we would consider as nonerror variance. Examine the matrix shown in Table 15.5. This is the same correlation matrix as shown in Table 15.3, except "1.0" has been entered where each variable intersects itself. These entries indicate that each variable is perfectly correlated with itself—there is no error variance (as was assumed in the example where this matrix was factored). If we made this assumption in running a factor analysis, we would incorporate such entries in the correlation matrix used as input for the analysis. (This is what is meant by "putting ones in the diagonals.") On the other hand, if we have some basis for estimating what this entry might be, some figure of less than 1.0 is located in the matrix for each variable. As mentioned earlier, such entries might be based on prior research. There are, however, a number of mathematical procedures for estimating these entries on the basis of characteristics of the variable's measurement distribution.

Given, then, assumptions about the measurement error attributable to each variable in an analysis, it remains for the solution itself to allocate the remaining variance on each variable (that is, $V_t - V_e$) in terms of that which is attributable to factors (V_c) and that which remains as specific to the variables (V_s).

Solution Model

Without getting into computational details, it is difficult to say much about the different types of factor analysis solutions. When we consider a particular solution model, this is a decision somewhat in addition to how measurement error has been defined and what rotational procedures will be undertaken. The term "somewhat" is used because some solutions are based on certain assumptions concerning measurement error, and some solutions carry particular implications concerning rotation or even the need for rotation. The term "in addition" is used because the solution itself centers on how factors are basically extracted from the correlation matrix, whereas the estimates of measurement error and rotational procedures involve additional decisions. Perhaps the most frequently en-

Table 15.5   Example of the Intercorrelation Matrix from Table 15.3 but Assumption of No Error Variance Is Indicated by Including "1.0" Entries in the Diagonals

	A	B	C	D	E
A	(1.0)	.985	.500	.342	.174
B		(1.0)	.642	.500	.342
C			(1.0)	.985	.940
D				(1.0)	.985
E					(1.0)

countered solution is one known as the *principal components* method. Suffice it to say that the chief feature of this method is that it attempts to extract a maximum amount of variance as each factor is calculated; it attempts to redefine the correlational matrix with as few factors as possible. Oftentimes, one will encounter this method, combined with an assumption of no measurement error (putting ones in the diagonals), and with use of the varimax rotational method.

Interpretation of the Rotated Factor Matrix

A practical point is to reemphasize that the interpretation of factors is a subjective decision on the part of the researcher. Nothing in the logic of a factor analysis solution tells a researcher how to label a particular factor. Interpretation of a factor involves examining what variables appear to load relatively high on that factor (and perhaps low on all others), then asking oneself if there is some more basic and meaningful variable that may underlie these high loading variables. If this more basic variable can be given a meaningful name, we use this name to label the factor. Thus, in the example given early in the chapter, it was said that measures of recall, comprehension, and reasoning all seemed to index a basic variable that we might call "understanding."

Give a particular solution where the factors have been meaningfully labeled, it is often important to make an overall evaluation of how the factors together characterize some overall meaningful construct. For example, we said earlier that all of the factors (Table 15.2) seemed to characterize the basic variables relevant to the use of television and classroom activity to reduce sex-role stereotyping. We can then consider how each of the facets of this construct variously contributes to the overall picture in terms of how common variation (V_c) breaks down into variation due to each of the factors (that is, V_I, V_{II}, and so on). Finally, we can see how the construct itself fits into the totality of what we have measured. That is, we can consider how much of the total variation (V_t) is accommodated by the factor structure itself (V_c) and how much variation remains specific to the original variables (V_s) and to measurement error (V_e).

Factor Scores

Often, when a researcher has a further analysis that follows the results of a factor analysis, *factor scores* are used. Such scores are simply a combination of variables that loaded on a factor. Thus, for example, suppose in the sample study in this chapter that we wished to compare boys and girls in terms of their overall understanding of the television materials. Recall that the factor analysis yielded a factor (I) that reflected primarily the combination of the individual measures of recall of television program content (Variable 4), comprehension of objectives (Variable 5), and reasoning from objectives (Variable 6). In this case, we might add or average these three variables together to form one composite variable for the comparison of boys' and girls' understanding. This combination could also be calculated by taking the variable loadings as a basis for "weighting" their contribution to the overall score. Such combinations, which may take a variety of forms, are generally known as *factor scores*. Many computer programs provide options for calculating factor scores as the weighted combination of all variables loading on a factor.

Confirmatory Factor Analysis

Most applications of factor analysis, including the example in this chapter, we considered to be "exploratory" or "descriptive" rather than a formal hypothesis test. As anybody who has used factor analysis can testify, decisions about scale reliability, rotational models, or especially factor identification often leave open many alternative solutions and interpretations. In recent years, an alternative approach called *confirmatory factor analysis* has been coming into vogue. A fundamental requirement for its use is that the researcher have grounds for making predictions about a given factor structure in the behavior of a specified population. Data are gathered from a sample, and an analysis is conducted to test the presence of the hypothesized structure. This, in turn, provides a basis for making an inference about the specified population based upon data from the sample.

For example, confirmatory factor analysis might be applied to the study described earlier in this chapter to test the presence of the three-factor model ("comprehension/exposure," "stereotyping/student's liking of materials," and "teacher's attitude/student's liking of materials") in a different population of students who saw the television series.

EXAMPLE FROM THE DEMOGRAPHICS AND MEDIA USES DATA

In Chapter 14, an example study was introduced in which demographic and media use measures had been obtained from a hypothetical sample of 101 individuals. Demographic variables included age, sex, education, and income. Media use variables were number of hours weekly of magazine reading, book reading, newspaper reading, watching broadcast television, watching VCR programs, and whether or not the subjects used a personal computer. The descriptive statistics and intercorrelation matrixes of these variables can be reviewed in Table 14.1 and 14.2, respectively (pp. 172, 173). You may recall that cluster analyses revealed that the sample had a few highly homogeneous, small clusters of individuals distinctive in income, education, and related media use, with the remainder of the group mostly undifferentiated in terms of major contrasts. What would factor analysis add to this study?

Results of a principal components analysis of these data are presented in Table 15.6.

Briefly, we can see that Factor I (perhaps labeled "demographics/media use") is bipolar, reflecting education, income, and print use at one end

Table 15.6   Principal Components Analysis of Demographic and Media Use Data from the Study Described in Chapter 14

Variable	Factor			
	I	II	III	IV
Age (years)	.40	.78	−.13	−.15
Sex (female = 0; male = 1)	−.08	.16	.22	.71
Education (years)	.87	.09	.09	−.01
Income (yearly)	.42	.23	.65	.14
Magazine use (hours weekly)	.66	.35	−.46	−.03
Book use (hours weekly)	.55	.31	−.54	.11
Newspaper use (hours weekly)	.25	.22	.55	.06
Broadcast television (hours weekly)	−.58	−.03	−.18	.08
VCR television (hours weekly)	.27	−.26	.23	−.69
PC use (0 = no; 1 = yes)	.35	−.71	−.24	.33

and television use at the other. This is simply another way of viewing what we would expect from the intercorrelations among these variables as earlier shown in Table 14.2. It also reflects some of the distinctions of the first clusters identified in the cluster analysis (Fig. 14.1, p. 174).

Factor II ("age/PC use") is also bipolar, reflecting the inverse relation between age and use of a personal computer.

Factor III ("income/type of print use") ferrets out distinctions within the use of print media, indicating that people higher in income (not education) tend to favor newspapers over books and magazines. Thus, although there was generally a pattern of higher social status (education, income) correlating with use of print over television, within the domain of print use, income as a separate variable somewhat differentiated uses of the different print media.

Factor IV ("sex/VCR use"), again bipolar, focuses on the relation of sex and VCR use, indicating that males as compared with females tend to make lesser use of VCRs.

In sum, factor analysis can "pull out" slightly more basic dimensions of relationships among the 10 variables, reducing them to four dimensions or factors. If you recall the types of clusters identified in the first few steps in Chapter 14, most of them reflected the types of multiple-variable relations shown here.

SUMMARY

From a practical standpoint, *factor analysis* is a method by which the relations among a relatively large number of variables are redefined in terms of relations with relatively fewer hypothetical variables called *factors*. Factor analysis procedure begins with a matrix of intercorrelations and estimates of the measurement error involved in each variable. The solution itself redefines the correlational matrix in terms of a *factor matrix*. There are then different procedures for *rotating* the factor matrix according to such criteria as *simple structure* or *psychological fit*. Given a rotated factor matrix, the researcher then attempts a subjective interpretation of the factors that have been obtained. Sources of variation involved in a factor analysis solution include definition of *common factor variance* (variance attributable to the factors extracted), *specific variance* (nonerror variance not included in the factor structure), and *error variance*. Among the practical aspects of factor analysis is the use of *confirmatory analysis* procedures to test hypotheses of factor structures based upon data from individuals sampled from a specified population.

SUPPLEMENTARY READINGS

Fink, Edward L., and Peter Monge, "An Exploration of Confirmatory Factor Analysis," in Brenda Dervin and M. Voight (eds.), *Progress in Communication Sciences*. Norwood, N.J.: Ablex, 1985.

Harman, H. H., *Modern Factor Analysis*. 3rd ed. Chicago: University of Chicago Press, 1976.

Hunter, John E., "Factor Analysis," Chapter 8 in Peter Monge and J. Cappella (eds.), *Multivariate Techniques in Communication Research*. New York: Academic Press, 1978.

Kim, J. O., and C. W. Mueller, *Introduction to Factor Analysis: What It Is and How to Do It*. Paper series on Quantitative Applications in the Social Sciences. Beverly Hills, Calif.: Sage, 1978.

SAS Institute, Inc., *SAS User's Guide: Statistics*. 1982 ed. Cary, N.C.: SAS Institute, 1982. See Chapters 21 (pp. 309–345) and 22 (347–367).

SPSS, Inc., *SPSS-X User's Guide*. 3rd ed. Chicago: SPSS, 1988. Chapter 28 describes a program for factor analysis.

16

Discriminant Analysis

Just as we can try to cluster individuals according to multiple measures (Chapter 14) or determine how these measures might be reduced to fewer, more basic, or underlying factors (Chapter 15), we can use discriminant analysis to determine how a linear combination of the multiple variables might differentiate individuals into some given group of categories. Applications of discriminant analysis are usually one of two types. One is to determine the degree to which members of different groups (for example, successful compared with unsuccessful teachers) can be differentiated in terms of an array of discriminator variables (for example, amount of training, time in service, size of school system). Another use is to classify new individuals or phenomena into categories, given the results of a previous analysis of the relationship between discriminating variables and categories (for example, predicting whether teachers are likely to be successful or unsuccessful).

A SAMPLE STUDY

As an example of the first type of use mentioned above, let us consider again several of the variables in the sample study of the use of educational television to combat sex-role stereotyping in children's judgments of occupations.

Problem

Recall that in the earlier description of the educational television series (Chap. 15, pp. 176–178) children had been exposed to a series of television programs and classroom activities which tended to reduce sex-role stereotyping. Subsequently, the children were tested on the degree to which they imposed male or female stereotypes when estimating who would fill a variety of occupations. This provided a male stereotype score (Variable 2; Table 15.1) and a female one (Variable 3). Also, the students answered multiple-choice test questions over the recall of television program content (Variable 4) and understanding of instructional objectives (Variable 6). As we have seen in examples of actor analysis there were differences in children's stereotyping and comprehension scores and these differences were variously related to programs they had seen, the number of classroom activities engaged in, and their teachers' attitudes toward the materials. For the present problem, the researcher is concerned with the degree to which the effects on stereotyping and comprehension scores are generalizable across students from the fourth, fifth, and sixth grades who were involved in the study. There is reason to believe, for example, that the older students may be more fixed in their stereotyped attitudes than the younger ones and, thus, may appear different in terms of the stereotype scores. Also, the producers of the television series and activities are concerned with the degree to which they successfully gauge the materials for children in this age group, having anticipated that most of what they presented was comprehensible to fourth through sixth graders. Finally, because the results of the project are in both the areas of attitudes and comprehension effects, it is important to determine the degree to which these two variables together might be different among the three groups.

The research question is: To what degree and how might the stereotyping scores and comprehension scores discriminate among fourth, fifth, and sixth graders in the project?

Method

It may be recalled that in the factor analysis (Chapter 15, pp. 180–181), it was found useful to combine the male and female stereotyping scores into a single *stereotyping* variable and the measures of recall, comprehension, and reasoning into one *understanding* variable. These two overall variables are then entered as the *discriminating* variables in the analysis. The *groups* variable in the analysis is whether a student was in fourth, fifth, or sixth grade.

A multiple discriminant analysis is undertaken, and it yields two *discriminant functions*. (Generally, the maximum number of discriminant functions is one less than the number of groups being compared.) Each of the discriminant functions represents a mathematical attempt to maximize a linear combination of the discriminating variables with the group variables. The first function represents the initial maximum linear configuration for discrimination of the dependent variable, then the second function in this case maximizes whatever is not included in the first function, and so on. For each function, calculations will yield: (1) a *canonical correlation coefficient*,[1] which describes the relationship of the discriminating variables with the groups variable; (2) the *standardized discriminant function coefficients*, which describe in standard score form the weightings of each of the discriminant variables on each of the functions; (3) the *unstandardized discriminant function coefficients*, which allow us to calculate a discriminant score by application to our raw measures; and (4) the *centroids*, or essentially the mean discriminant scores for each group on each function.[2]

---

[1] Canonical correlation indexes the degree of correlation between the linear combination of two sets of variables rather than individual variables. Somewhat akin to linear regression analysis, the canonical analysis seeks the maximum linear relationship between linear combinations for each set of variables. The canonical coefficient is interpreted in a manner similar to a coefficient of correlation.

[2] Other results are provided by computer programs for multiple discriminant analysis, such as *Wilks' lambda* calculations used to determine whether amounts of variance accounted for by discriminant variables are significant as well as several other calculations that are beyond the scope of this discussion.

Results

For the sample problem, the multiple discriminant calculations reveal that the canonical correlation between the first discriminant function and the groups is $+.728$. This is a moderately high correlation and statistically significant ($p < .01$). By contrast, the canonical correlation between the second discriminant function and the groups is $+.143$, which is low and not significant. Because of this, we shall drop the second discriminant function from further practical interpretation. We know that the groups are discriminated chiefly on the first function.

What is the makeup of the first function? The standardized discriminant function coefficients, which tell us this, are nearly 1 for the *understanding* variable, and nearly 0 for the *stereotyping* variable. The interpretation is that, for all practical purposes, the first discriminant function mainly reflects scores on the understanding variable. Thus, the answer to the basic question is that the fourth-, fifth-, and sixth-grade groups are significantly discriminated on the understanding measures and not on the stereotyping ones.

Unstandardized discriminant function coefficients allow the researcher to add raw scores on the discriminating variables to a constant and, then, convert them into discriminant scores. (If we first standardize our measures, then the standardized discriminant function coefficients may be used for this purpose.) Discriminant scores are standardized with the mean of all data across the groups equal to 0, and the standard deviation equal to one. This allows us to interpret the distribution of discriminant scores in terms of probabilities. In the present results, the unstandardized coefficients are 0.58986 for the understanding score, 0.02519 for the stereotyping score, and -2.70259 for the constant. Thus, for example, if the mean raw scores of the fourth-grade group were 2.75 on understanding and 5.25 on stereotyping, these multiplied by their respective unstandardized coefficients and summed with the constant will equal -0.95259, or the discriminant score for that group on the first discriminant function.

Recall that a *centroid* is the mean discriminant score on a function for a group. The mean discriminant score that we just calculated for the fourth graders is a centroid. For the other groups, the centroids on the first function are 0.13633 for fifth graders and 0.81190 for sixth graders. From these, we can see that understanding increases across the grades.

We can also use multiple discriminant analysis results to compare a new case with prior classifications. Suppose, for example, that a new fifth-grade class went through the same study and its understanding score was 4.8 and its stereotyping score was 5.0. Based on prior results, would they fit in with the other fifth graders? We can answer this question by multiplying these raw scores by their respective unstandardized coefficients, then summing the results with the constant to yield a discriminant score. The calculated score is $+0.25741$, which is closest to the centroid of the earlier fifth graders ($+0.13633$); hence, the new fifth graders would be classified into the same grade in terms of their performances.

Let us, however, look at another sample study where classification is the main objective.

A SECOND SAMPLE STUDY

As mentioned at the outset of this chapter, a second use of multiple discriminant analysis is in assigning previously unclassified individuals or phenomena to groups, based on results of a multiple discriminant analysis where both group assignment and the relation with the discriminant variables are known. This second sample study is also drawn from the earlier educational television sample study.

Problem

As previously described in Chapter 15, we know that children's understanding of the television materials devoted to reduction of sex-role stereotyping was predictable on the basis of the number of television programs they had viewed and the number of classroom activities in which they had engaged in on the topic. For the present example, let us consider that an overall score of 4.0 or greater represents a *suitable level* of understanding of the materials, whereas a score lower than this represents an *unsuitable level*. Given the data we already have on children, it is possible to undertake a multiple discriminant analysis of the relationship between discriminant variables of television-viewing frequency and activities frequency with the group classification of having a suitable or unsuitable level of understanding of the materials. Results of the analysis could be a basis for classifying children into likely (1) suitable or (2) unsuitable comprehension groups based on a knowledge of the number of television programs they have viewed and activities they have engaged in, both relevant to the reduction of sex-role stereotyping.

Method

A discriminant analysis is undertaken where the discriminating variables are television-viewing frequency (Variable 8, Table 15.1) and activities frequency (Variable 9). The groups variable is the separation of children with a suitable compared with an unsuitable level of understanding of the materials.

Results

For the sample problem, the multiple discrimination analysis reveals one discriminant function with an associated canonical correlation of $+.774$. This is a relatively high degree of correlation between the discriminating variables and the groups.

Furthermore, the standardized discriminant function coefficients are

.781 for television-viewing frequency and .492 for the classroom activities. It is clear that both variables contribute to differentiating groups of children in terms of suitable compared with unsuitable levels of understanding of the materials. However, television-viewing frequency contributes substantially more than does the activities frequency.

The unstandardized discriminant coefficients are 0.41110 for television-viewing frequency and 0.17496 for activities frequency with a constant of −2.76479. If the mean television-viewing frequency of the group having an *unsuitable* level of understanding is 3.40 and the activities mean is 2.60, then (applying the above) its centroid is −0.91217. Similarly, with a mean viewing frequency of 6.0 and activities mean of 5.4286, the centroid of the *suitable* level of understanding for the group is 0.65155. The midpoint between these two values is −0.12845.

Consider the continuum below for discriminant scores in this problem:

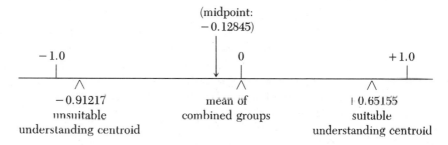

Again, the discriminant score is standardized; its mean equals 0, and this is the mean of the discriminant scores of both groups combined. Note relatively where the centroids of the two groups fall on this continuum. One practical way to classify new cases is on the proximity of their discriminant scores to the two group centroids. If, for example, a new child had a television-viewing frequency of four programs and had undergone class activities four times, the discriminant score would be:

$$(4)(.4111) + (4)(.17496) + (-2.76479)$$
$$\text{(or)}$$
$$-.42479$$

The new discriminant score is closer to the unsuitable understanding group centroid (to the left of −0.12845 above). This leads to the classification of the new case into that group.

NOTES ON MULTIPLE DISCRIMINANT ANALYSIS

Relation with Other Methods

If you have read or even browsed in the earlier chapters on regression analysis or factor analysis, you should see a general conceptual relation of all of these methods with multiple discriminant analysis. As was said at the outset of this chapter, when the discrimination involves only two groups, multiple discriminant analysis and multiple regression analysis are essentially the same, except that the multiple discriminant analysis has a binary or dummy dependent variable in this case. In linear regression, we are asking: What is the maximum combined linear combination of a cluster of independent variables with a dependent variable? In canonical correlation, we are asking: What is the maximum linear relationship, or relationships, between two clusters of variables? In discriminant analysis, we are asking: What is the maximum linear relationship, or relationships, of a cluster of variables with a variable that is divided into categories?

As in all the statistical methods ranging from analysis of variance through correlation, regression, and factor analysis, we are trying to account for variance reflected in relationships among variables. In multiple discriminant analysis, as in factor analysis and canonical correlation, we are *transforming* the known relationships among individual variables into another configuration. In factor analysis, the larger array of relationships is reduced to fewer clusters of relationships called *factors*. In canonical correlation, clusters are interrelated in terms of *canonical variates*. In multiple discriminant analysis, the relationship is characterized as *discriminant functions*. As in regression analysis, factor analysis, or canonical correlation, we can attempt to interpret these clusters in terms of coefficients that indicate the degree of relationship of each of our original variables with the cluster. In the case of multiple discriminant analysis, each discriminant function is interpreted in terms of the various weightings of the original variables with that function.

Although the aforementioned methods all have much in common, they all have their own biases in what they reveal to us. The factors in factor analysis will, for the most part, be biased toward taking into account as much of the common variance in an overall matrix of interrelationships as is possible under the terms of the particular solution. In canonical correlation, the canonical variates will be biased in definition toward the maximum linear relationships of two clusters of variables. As was said in the previous chapter, although the researcher can always see some similarities between canonical variates and factors for the same set of data, these similarities will not necessarily be identical. A discriminant function, like a canonical variate or a factor, may reflect different aspects of the same patterns of common variance among an array of variables but will not necessarily have identical patterns with them.

Interpretation of Coefficients

Again, both standardized and unstandardized discriminant function coefficients are useful in interpretation of discriminant functions as well as discriminant scores.

Standardized coefficients, because they equalize measurement scales, are the basis for interpreting relative contributions of discriminant variables to a function. They can also be used to convert discriminating variable measures to discriminant scores *if* the former are first standardized (which they usually are not).

Unstandardized coefficients and the constant, like regression coefficients, are a basis for calculating discriminant scores from raw (unstandardized) scores of the discriminating variables. Because they also reflect measurement-scale differences, they are not an accurate basis for assessing relative contribution of discriminating variables to a discriminant function.

Particularly when multiple discriminant analysis is being used to classify new cases, based on known relationships between discriminating variables and groups, the equation involving unstandardized coefficients can be used in practical ways to show the overall relation as well as a practical guide for making predictions. The classification chart in Table 16.1 illustrates this for the second sample study.

Given the number of television programs viewed and the activities frequency as a basis for entering the chart, we can apply the coefficients and calculate likely group classification (S = satisfactory understanding, U = unsatisfactory). Note how the overall pattern in Table 16.1 reflects that television program exposure contributes more to gaining satisfactory understanding (S) than does activities frequency. If we were using this table for practical scheduling of programs, we would know that scheduling fewer than seven programs without activities would be unsatisfactory. Perhaps optimal is the combination of five programs and four or five activities sessions. Again, this table is one way the relations between the discriminating variables and the groups, as defined by the unstandardized coefficients, can be illustrated.

Table 16.1 A Classification Guide for Predicting Likely Satisfactory (S) or Unsatisfactory (U) Understanding of Sex-Role Stereotyping Materials Given Television-Viewing and Class Activities Frequencies (Simulated Data)

		0	1	2	3	4	5	6	7	8	9	10	11	12	13	14	15	16
TV-	7	S	S	S	S	S	S	S	S									
Viewing	6	U	S	S	S	S	S	S	S									
Frequency	5	U	U	U	U	S	S	S	S									
	4	U	U	U	U	U	U	U	S	S	S	S	S	S	S	S		
	3	U	U	U	U	U	U	U	U	U	S	S	S	S	S	S		
	2	U	U	U	U	U	U	U	U	U	U	U	U	S	S	S	S	S
	1	U	U	U	U	U	U	U	U	U	U	U	U	U	U	S	S	S
	0	U	U	U	U	U	U	U	U	U	U	U	U	U	U	U	U	S
		0	1	2	3	4	5	6	7	8	9	10	11	12	13	14	15	16

Class Activities Frequency

Discriminant Scores and Probabilities

It is useful to remember that these are standardized scores with a mean of 0 and a standard deviation of 1. The 0 is the standard score representation of the overall mean of the combined groups on the discriminant score on that function. Thus, if a centroid of one group is -1.0, we can estimate that 34 percent of the cases in the overall distribution are apt to fall between the mean of that group and the mean of the overall distribution. If the centroid of a second group is $+0.50$, then we know that 53 percent of the cases are likely to fall *between* the means of the two groups on that discriminant function. Or for that matter, if cases in a group are between -1.8 and -0.05 on a function, then 44 percent of the cases in the overall distribution are likely to fall in that group on that function.

Obviously, when the correlation of a discriminant function is not perfect (that is, canonical correlation $\neq 1.0$), some error will be reflected in the assignment of cases to groups. This can be considered error in the body of data under analysis or likely error in assigning new cases. In essence, "error" occurs when a case that prior data show belonging to one group is, on the basis of the equation, assigned to another. Error is also reflected in the "overlap" of likely distributions around different centroids; the higher the probability of overlap, the higher is the likely error. Packaged computer programs for multiple discriminant analysis sometimes provide subroutines that will plot distributions and their overlaps. In some practical applications of multiple discriminant analysis to make classifications, the researcher may wish to designate a zone of "unclassified" cases where the probability of erroneous classification is high (say, as shown in a standard score plot).

It is also possible to calculate the assignment of cases to groups, based on the unstandardized coefficients, then to compare this with the known group assignment in the input data. This is illustrated in Table 16.2 for the second sample study. Note how more cases in the "satisfactory" understanding group are likely to be misclassified. This is because the distribution of this group overlaps more into the lower group rather than

Table 16.2   Classification of Students into "Satisfactory" and "Unsatisfactory" Understanding Groups of Sex-Role Stereotyping Based on Multiple Discriminant Analysis of Television-Viewing and Class Activities Frequencies (Simulated Data)

Known Group	Predicted Group	
	Unsatisfactory Understanding	Satisfactory Understanding
Unsatisfactory ($N = 50$) understanding	50	0
Satisfactory ($N = 70$) understanding	10	60

vice versa. (This was illustrated somewhat in the continuum on p. 199. Note how the centroid of the "satisfactory" group is closer to the overall mean or 0 and how the dividing point is to the left of this overall mean.) Of the 120 cases, 110, or approximately 92 percent, are assigned correctly. This is a very practical way of considering "error," both in the known data or in likely classification of new cases.

EXAMPLE FROM THE DEMOGRAPHICS AND MEDIA USES DATA

Again, in Chapter 14, an example study was introduced in which demographic and media use measures had been obtained from a hypothetical sample of 101 individuals. Demographic and media variables included age, sex, education, income, number of hours weekly of magazine reading, book reading, newspaper reading, watching broadcast television, watching VCR programs, and whether or not an individual used a personal computer. The descriptive statistics and intercorrelation matrixes of these variables can be reviewed in Tables 14.1 and 14.2, respectively (pp. 172, 173). You may recall that cluster analyses revealed that the sample had a few highly homeogeneous small clusters of individuals distinctive in income, education, and related media use, with the remainder of the group mostly undifferentiated in terms of major contrasts. Also, as described in Chapter 15, a principal components (factor) analysis of these data revealed that the 10 variables could be reduced to four factors—"demographics/media use," "age/PC use," "income/type of print use," and "sex/VCR use."

Now, in the context of the present chapter, how could discriminant analysis be applied to these data? For an example, let us examine the degree to which the measures might, as a linear combination, best discriminate individuals' uses of personal computers.

Such an analysis was calculated with the following results:

The canonical correlation between the first discriminant function and the PC user/nonuser groups is .663, which is significant at the $p < .01$ level.

The variables contributing most to the above correlation are age (standardized function coefficient $= -.98$), education (.66), and TV viewing (.45). (For ease of interpretation, the signs have been reversed from the actual analysis, but the meanings are the same.)

The canonical correlation between the second discriminant function and the PC user/nonuser groups is .605, which is also significant at the $p < .01$ level.

The variables contributing most to the second correlation are age (standardized function coefficient $= -.92$), education (.34), and income $(-.36)$.

In all, age is by far the strongest predictor of personal computer use, older people not using one. Younger, educated individuals but without necessarily a high income and who watch television (which, apart from education, seems to accompany income) are the personal computer users. Again, the results reflect the major bases of distinguishing many of the clusters (Chapter 14), the interrelations among the variables (like "factors," Chapter 15), and altogether some of the overall relations between demographics and media uses.

SUMMARY

Multiple discriminant analysis provides for the differentiation of single-variable groups or categories on the basis of relations with an array of discriminating variables. The calculation seeks to identify maximum multiple linear relations of discriminating variables with the groups. Each set of relations is called a *discriminant function*. For each discriminant function, the calculations also provide (1) *the canonical correlation* of that function with the groups; (2) *standardized discriminant function coefficients*, which reflect the relative contributions of discriminating variables to that function; (3) *unstandardized discriminant function coefficients* and a constant, which may be used to convert raw scores on discriminating variables to a discriminant function score; and (4) *centroids*, or the average discriminant function scores for each of the groups. Multiple discriminant analysis is used both to describe the differentiation of groups, based on discriminating variables, and is used as a basis for classifying new cases into likely groups. Such analyses are part of a family of multivariate methods, all of which involve assessments of relations within, among, or between groups of variables. Discriminant function scores are standardized and are, thus, interpretable in terms of distributional probabilities. A practical gauge of error in multiple discriminant analysis is the comparison of prediction of group assignments with known ones.

SUPPLEMENTARY READINGS

Klecka, W. R., *Discriminant Analysis*. Paper series on Quantitative Application in the Social Sciences. Beverly Hills, Calif.: Sage, 1980.

McLaughlin, Margaret L., "Discriminant Analysis in Communication Research," Chapter 6 in Peter Monge and J. Cappella (eds.), *Multivariate Techniques in Communication Research*. New York: Academic Press, 1978.

Marasculio, Leonard A., and Joel R. Levin, *Multivariate Statistics in the Social Sciences*. Monterey, Calif.: Brooks/Cole, 1983. Chapters 6 and 7 describe discriminant analysis.

SAS Institute, Inc., *SAS User's Guide: Statistics*. 1982 ed. Cary, N.C.: SAS Institute, 1982. See the section in "Discriminant Analysis," pp. 365–414.

SPSS, Inc., *SPSS-X User's Guide*. 3rd ed. Chicago: SPSS, 1988. Chapter 27 describes a program for discriminant analysis.

TABLES

A. PERCENTAGE POINTS OF THE t DISTRIBUTION
B. PERCENTAGE POINTS OF THE F DISTRIBUTION
C. PERCENTAGE POINTS OF THE χ^2 DISTRIBUTION

Use of the t Table

In Table A, v is degrees of freedom ($d.f.$). When testing the difference between the means of two independent groups (Chapter 7), $v = n_I + n_{II} - 2$; for single or matched groups, $v = n - 1$. Enter the row corresponding to the value of v for the problem and locate the value of t that was calculated. The column headings provide the associated probability values (p). Probability values for one-tailed tests are those listed for Q; two-tailed test probabilities are those listed for $2Q$.

In practice, t will usually fall between two columns. Customarily, we are interested in whether t falls in a position indicating $p < .05$ or $p < .01$, or whatever the significance level might be. For example, for a two-tailed test, if $n = 14$ and $t = 2.150$, then probability could be reported as $p < .05$. Sometimes it is desirable to report the probability values whose columns bracket the t value. Thus, in the present case, we might report $.02 < p < .05$, which is the same as saying, "probability greater than .02 but less than .05." If the calculated value of t exceeds the value shown in the column on the far right in Table A, it would be customary to report $p < .005$ for a one-tailed test, or $p < .01$ for a two-tailed test.

Use of the F Table

In Table B, v_1 is degrees of freedom ($d.f.$) in the numertor of the F ratio, and v_2 is degrees of freedom in the denominator. Determination of degrees of freedom for F ratios is discussed in Chapters 8 and 9. Table B is divided into two sections: "Upper 5 percent points" refers to values of F for significance at the .05 level; "Upper 1 percent points" refers to values for the .01 level. Depending on what significance level has been set, one of these tables is selected.

The table is entered at the intersection of the row (v_2) and column (v_1) corresponding to the degrees of freedom for a given F ratio. If the calculated value of F is equal to or exceeds the tabular entry, then $p < .05$ or $p < .01$, depending on which section of the table is being used. For example, if the significance level has been set at .05, and $v_1 = 3$, $v_2 = 9$, we would interpret an F of 3.90 as statistically significant ($p < .05$).

Table A. Percentage Points of the t Distribution

ν	$Q = 0.4$ $2Q = 0.8$	0.25 0.5	0.1 0.2	0.05 0.1	0.025 0.05	0.01 0.02	0.005 0.01
1	0.325	1.000	3.078	6.314	12.706	31.821	63.657
2	.289	0.816	1.886	2.920	4.303	6.965	9.925
3	.277	.765	1.638	2.353	3.182	4.541	5.841
4	.271	.741	1.533	2.132	2.776	3.747	4.604
5	0.267	0.727	1.476	2.015	2.571	3.365	4.032
6	.265	.718	1.440	1.943	2.447	3.143	3.707
7	.263	.711	1.415	1.895	2.365	2.998	3.499
8	.262	.706	1.397	1.860	2.306	2.896	3.355
9	.261	.703	1.383	1.833	2.262	2.821	3.250
10	0.260	0.700	1.372	1.812	2.228	2.764	3.169
11	.260	.697	1.363	1.796	2.201	2.718	3.106
12	.259	.695	1.356	1.782	2.179	2.681	3.055
13	.259	.694	1.350	1.771	2.160	2.650	3.012
14	.258	.692	1.345	1.761	2.145	2.624	2.977
15	0.258	0.691	1.341	1.753	2.131	2.602	2.947
16	.258	.690	1.337	1.746	2.120	2.583	2.921
17	.257	.689	1.333	1.740	2.110	2.567	2.898
18	.257	.688	1.330	1.734	2.101	2.552	2.878
19	.257	.688	1.328	1.729	2.093	2.539	2.861
20	0.257	0.687	1.325	1.725	2.086	2.528	2.845
21	.257	.686	1.323	1.721	2.080	2.518	2.831
22	.256	.686	1.321	1.717	2.074	2.508	2.819
23	.256	.685	1.319	1.714	2.069	2.500	2.807
24	.256	.685	1.318	1.711	2.064	2.492	2.797
25	0.256	0.684	1.316	1.708	2.060	2.485	2.787
26	.256	.684	1.315	1.706	2.056	2.479	2.779
27	.256	.684	1.314	1.703	2.052	2.473	2.771
28	.256	.683	1.313	1.701	2.048	2.467	2.763
29	.256	.683	1.311	1.699	2.045	2.462	2.756
30	0.256	0.683	1.310	1.697	2.042	2.457	2.750
40	.255	.681	1.303	1.684	2.021	2.423	2.704
60	.254	.679	1.296	1.671	2.000	2.390	2.660
120	.254	.677	1.289	1.658	1.980	2.358	2.617
∞	.253	.674	1.282	1.645	1.960	2.326	2.576

This table is abridged from Table 12 in the *Biometrika Tables for Statisticians,* Vol. 1, Third edition (1966), E. S. Pearson and H. O. Hartley (Eds.). Reproduced by permission of the Biometrika Trustees.

Columns headed Q = 0.1 and 0.01 in Table A are taken from Table III of Fisher and Yates: *Statistical Tables for Biological, Agricultural and Medical Research,* published by Longman Group Ltd., London. (Previously published by Oliver and Boyd, Edinburgh), and by permission of the authors and publishers.

Table B. Percentage Points of the *F* Distribution Upper 5 Percent Points

ν_2 \ ν_1	1	2	3	4	5	6	7	8	9	10	12	15	120	∞
1	161.4	199.5	215.7	224.6	230.2	234.0	236.8	238.9	240.5	241.9	243.9	245.9	253.3	254.3
2	18.51	19.00	19.16	19.25	19.30	19.33	19.35	19.37	19.38	19.40	19.41	19.43	19.49	19.50
3	10.13	9.55	9.28	9.12	9.01	8.94	8.89	8.85	8.81	8.79	8.74	8.70	8.55	8.53
4	7.71	6.94	6.59	6.39	6.26	6.16	6.09	6.04	6.00	5.96	5.91	5.86	5.66	5.63
5	6.61	5.79	5.41	5.19	5.05	4.95	4.88	4.82	4.77	4.74	4.68	4.62	4.40	4.36
6	5.99	5.14	4.76	4.53	4.39	4.28	4.21	4.15	4.10	4.06	4.00	3.94	3.70	3.67
7	5.59	4.74	4.35	4.12	3.97	3.87	3.79	3.73	3.68	3.64	3.57	3.51	3.27	3.23
8	5.32	4.46	4.07	3.84	3.69	3.58	3.50	3.44	3.39	3.35	3.28	3.22	2.97	2.93
9	5.12	4.26	3.86	3.63	3.48	3.37	3.29	3.23	3.18	3.14	3.07	3.01	2.75	2.71
10	4.96	4.10	3.71	3.48	3.33	3.22	3.14	3.07	3.02	2.98	2.91	2.85	2.58	2.54
11	4.84	3.98	3.59	3.36	3.20	3.09	3.01	2.95	2.90	2.85	2.79	2.72	2.45	2.40
12	4.75	3.89	3.49	3.26	3.11	3.00	2.91	2.85	2.80	2.75	2.69	2.62	2.34	2.30
13	4.67	3.81	3.41	3.18	3.03	2.92	2.83	2.77	2.71	2.67	2.60	2.53	2.25	2.21
14	4.60	3.74	3.34	3.11	2.96	2.85	2.76	2.70	2.65	2.60	2.53	2.46	2.18	2.13
15	4.54	3.68	3.29	3.06	2.90	2.79	2.71	2.64	2.59	2.54	2.48	2.40	2.11	2.07
16	4.49	3.63	3.24	3.01	2.85	2.74	2.66	2.59	2.54	2.49	2.42	2.35	2.06	2.01
17	4.45	3.59	3.20	2.96	2.81	2.70	2.61	2.55	2.49	2.45	2.38	2.31	2.01	1.96
18	4.41	3.55	3.16	2.93	2.77	2.66	2.58	2.51	2.46	2.41	2.34	2.27	1.97	1.92
19	4.38	3.52	3.13	2.90	2.74	2.63	2.54	2.48	2.42	2.38	2.31	2.23	1.93	1.88
20	4.35	3.49	3.10	2.87	2.71	2.60	2.51	2.45	2.39	2.35	2.28	2.20	1.90	1.84
21	4.32	3.47	3.07	2.84	2.68	2.57	2.49	2.42	2.37	2.32	2.25	2.18	1.87	1.81
22	4.30	3.44	3.05	2.82	2.66	2.55	2.46	2.40	2.34	2.30	2.23	2.15	1.84	1.78
23	4.28	3.42	3.03	2.80	2.64	2.53	2.44	2.37	2.32	2.27	2.20	2.13	1.81	1.76
24	4.26	3.40	3.01	2.78	2.62	2.51	2.42	2.36	2.30	2.25	2.18	2.11	1.79	1.73
25	4.24	3.39	2.99	2.76	2.60	2.49	2.40	2.34	2.28	2.24	2.16	2.09	1.77	1.71
26	4.23	3.37	2.98	2.74	2.59	2.47	2.39	2.32	2.27	2.22	2.15	2.07	1.75	1.69
27	4.21	3.35	2.96	2.73	2.57	2.46	2.37	2.31	2.25	2.20	2.13	2.06	1.73	1.67
28	4.20	3.34	2.95	2.71	2.56	2.45	2.36	2.29	2.24	2.19	2.12	2.04	1.71	1.65
29	4.18	3.33	2.93	2.70	2.55	2.43	2.35	2.28	2.22	2.18	2.10	2.03	1.70	1.64
30	4.17	3.32	2.92	2.69	2.53	2.42	2.33	2.27	2.21	2.16	2.09	2.01	1.68	1.62
40	4.08	3.23	2.84	2.61	2.45	2.34	2.25	2.18	2.12	2.08	2.00	1.92	1.58	1.51
60	4.00	3.15	2.76	2.53	2.37	2.25	2.17	2.10	2.04	1.99	1.92	1.84	1.47	1.39
120	3.92	3.07	2.68	2.45	2.29	2.17	2.09	2.02	1.96	1.91	1.83	1.75	1.35	1.25
∞	3.84	3.00	2.60	2.37	2.21	2.10	2.01	1.94	1.88	1.83	1.75	1.67	1.22	1.00

This table is abridged from Table 18 in the *Biometrika Tables for Statisticians,* Vol. I, Third edition (1966), E. S. Pearson and H. O. Hartley (Eds.). Reproduced by permission of the Biometrika Trustees.

Table B. *(Cont.)* Percentage Points of the *F* Distribution Upper 1 Percent Points

v_2 \ v_1	1	2	3	4	5	6	7	8	9	10	12	15	120	∞
1	4052.	4999.5	5403.	5625.	5764.	5859.	5928.	5982.	6022.	6056.	6106.	6157.	6339.	6366.
2	98.50	99.00	99.17	99.25	99.30	99.33	99.36	99.37	99.39	99.40	99.42	99.43	99.49	99.50
3	34.12	30.82	29.46	28.71	28.24	27.91	27.67	27.49	27.35	27.23	27.05	26.87	26.22	26.13
4	21.20	18.00	16.69	15.98	15.52	15.21	14.98	14.80	14.66	14.55	14.37	14.20	13.56	13.46
5	16.26	13.27	12.06	11.39	10.97	10.67	10.46	10.29	10.16	10.05	9.89	9.72	9.11	9.02
6	13.75	10.92	9.78	9.15	8.75	8.47	8.26	8.10	7.98	7.87	7.72	7.56	6.97	6.88
7	12.25	9.55	8.45	7.85	7.46	7.19	6.99	6.84	6.72	6.62	6.47	6.31	5.74	5.65
8	11.26	8.65	7.59	7.01	6.63	6.37	6.18	6.03	5.91	5.81	5.67	5.52	4.95	4.86
9	10.56	8.02	6.99	6.42	6.06	5.80	5.61	5.47	5.35	5.26	5.11	4.96	4.40	4.31
10	10.04	7.56	6.55	5.99	5.64	5.39	5.20	5.06	4.94	4.85	4.71	4.56	4.00	3.91
11	9.65	7.21	6.22	5.67	5.32	5.07	4.89	4.74	4.63	4.54	4.40	4.25	3.69	3.60
12	9.33	6.93	5.95	5.41	5.06	4.82	4.64	4.50	4.39	4.30	4.16	4.01	3.45	3.36
13	9.07	6.70	5.74	5.21	4.86	4.62	4.44	4.30	4.19	4.10	3.96	3.82	3.25	3.17
14	8.86	6.51	5.56	5.04	4.69	4.46	4.28	4.14	4.03	3.94	3.80	3.66	3.09	3.00
15	8.68	6.36	5.42	4.89	4.56	4.32	4.14	4.00	3.89	3.80	3.67	3.52	2.96	2.87
16	8.53	6.23	5.29	4.77	4.44	4.20	4.03	3.89	3.78	3.69	3.55	3.41	2.84	2.75
17	8.40	6.11	5.18	4.67	4.34	4.10	3.93	3.79	3.68	3.59	3.46	3.31	2.75	2.65
18	8.29	6.01	5.09	4.58	4.25	4.01	3.84	3.71	3.60	3.51	3.37	3.23	2.66	2.57
19	8.18	5.93	5.01	4.50	4.17	3.94	3.77	3.63	3.52	3.43	3.30	3.15	2.58	2.49
20	8.10	5.85	4.94	4.43	4.10	3.87	3.70	3.56	3.46	3.37	3.23	3.09	2.52	2.42
21	8.02	5.78	4.87	4.37	4.04	3.81	3.64	3.51	3.40	3.31	3.17	3.03	2.46	2.36
22	7.95	5.72	4.82	4.31	3.99	3.76	3.59	3.45	3.35	3.26	3.12	2.98	2.40	2.31
23	7.88	5.66	4.76	4.26	3.94	3.71	3.54	3.41	3.30	3.21	3.07	2.93	2.35	2.26
24	7.82	5.61	4.72	4.22	3.90	3.67	3.50	3.36	3.26	3.17	3.03	2.89	2.31	2.21
25	7.77	5.57	4.68	4.18	3.85	3.63	3.46	3.32	3.22	3.13	2.99	2.85	2.27	2.17
26	7.72	5.53	4.64	4.14	3.82	3.59	3.42	3.29	3.18	3.09	2.96	2.81	2.23	2.13
27	7.68	5.49	4.60	4.11	3.78	3.56	3.39	3.26	3.15	3.06	2.93	2.78	2.20	2.10
28	7.64	5.45	4.57	4.07	3.75	3.53	3.36	3.23	3.12	3.03	2.90	2.75	2.17	2.06
29	7.60	5.42	4.54	4.04	3.73	3.50	3.33	3.20	3.09	3.00	2.87	2.73	2.14	2.03
30	7.56	5.39	4.51	4.02	3.70	3.47	3.30	3.17	3.07	2.98	2.84	2.70	2.11	2.01
40	7.31	5.18	4.31	3.83	3.51	3.29	3.12	2.99	2.89	2.80	2.66	2.52	1.92	1.80
60	7.08	4.98	4.13	3.65	3.34	3.12	2.95	2.82	2.72	2.63	2.50	2.35	1.73	1.60
120	6.85	4.79	3.95	3.48	3.17	2.96	2.79	2.66	2.56	2.47	2.34	2.19	1.53	1.38
∞	6.63	4.61	3.78	3.32	3.02	2.80	2.64	2.51	2.41	2.32	2.18	2.04	1.32	1.00

Use of the Chi-Square Table

In Table C, v is degrees of freedom $(d.f.)$, and Q stands for probability values (p). As discussed in Chapter 10, the one-sample case of chi-square analysis has v equal to the number of categories minus one; the multiple sample case has v equal to the number of rows minus one being compared multiplied by the number of columns minus one.

Enter the row corresponding to the value of v for the problem, and locate the column or position between columns where the calculated value of χ^2 falls. Probability values are then read from the column headings. For example, if $v = 4$, and $\chi^2 = 13.50$, we would report $p < .01$, or .005 $< p$ .01. If χ^2 is a value greater than the value in the .001 column, it would be customary to report $p < .001$.

Table C. Percentage Points of the χ^2 Distribution

$Q.$ / ν	0.250	0.100	0.050	0.025	0.010	0.005	0.001
1	1.32330	2.70554	3.84146	5.02389	6.63490	7.87944	10.828
2	2.77259	4.60517	5.99147	7.37776	9.21034	10.5966	13.816
3	4.10835	6.25139	7.81473	9.34840	11.3449	12.8381	16.266
4	5.38527	7.77944	9.48773	11.1433	13.2767	14.8602	18.467
5	6.62568	9.23635	11.0705	12.8325	15.0863	16.7496	20.515
6	7.84080	10.6446	12.5916	14.4494	16.8119	18.5476	22.458
7	9.03715	12.0170	14.0671	16.0128	18.4753	20.2777	24.322
8	10.2188	13.3616	15.5073	17.5346	20.0902	21.9550	26.125
9	11.3887	14.6837	16.9190	19.0228	21.6660	23.5893	27.877
10	12.5489	15.9871	18.3070	20.4831	23.2093	25.1882	29.588
11	13.7007	17.2750	19.6751	21.9200	24.7250	26.7569	31.264
12	14.8454	18.5494	21.0261	23.3367	26.2170	28.2995	32.909
13	15.9839	19.8119	22.3621	24.7356	27.6883	29.8194	34.528
14	17.1170	21.0642	23.6848	26.1190	29.1413	31.3193	36.123
15	18.2451	22.3072	24.9958	27.4884	30.5779	32.8013	37.697
16	19.3688	23.5418	26.2962	28.8454	31.9999	34.2672	39.252
17	20.4887	24.7690	27.5871	30.1910	33.4087	35.7185	40.790
18	21.6049	25.9894	28.8693	31.5264	34.8053	37.1564	42.312
19	22.7178	27.2036	30.1435	32.8523	36.1908	38.5822	43.820
20	23.8277	28.4120	31.4104	34.1696	37.5662	39.9968	45.315
21	24.9348	29.6151	32.6705	35.4789	38.9321	41.4010	46.797
22	26.0393	30.8133	33.9244	36.7807	40.2894	42.7956	48.268
23	27.1413	32.0069	35.1725	38.0757	41.6384	44.1813	49.728
24	28.2412	33.1963	36.4151	39.3641	42.9798	45.5585	51.179
25	29.3389	34.3816	37.6525	40.6465	44.3141	46.9278	52.620
26	30.4345	35.5631	38.8852	41.9232	45.6417	48.2899	54.052
27	31.5284	36.7412	40.1133	43.1944	46.9630	49.6449	55.476
28	32.6205	37.9159	41.3372	44.4607	48.2782	50.9933	56.892
29	33.7109	39.0875	42.5569	45.7222	49.5879	52.3356	58.302
30	34.7998	40.2560	43.7729	46.9792	50.8922	53.6720	59.703
40	45.6160	51.8050	55.7585	59.3417	63.6907	66.7659	73.402
50	56.3336	63.1671	67.5048	71.4202	76.1539	79.4900	86.661
60	66.9814	74.3970	79.0819	83.2976	88.3794	91.9517	99.607
70	77.5766	85.5271	90.5312	95.0231	100.425	104.215	112.317
80	88.1303	96.5782	101.879	106.629	112.329	116.321	124.839
90	98.6499	107.565	113.145	118.136	124.116	128.299	137.208
100	109.141	118.498	124.342	129.561	135.807	140.169	149.449

SIMPLIFIED COMPUTER PROGRAMS

MEAN, VARIANCE, AND STANDARD DEVIATION
t TEST
CHI SQUARE
CORRELATION
REGRESSION
DISTRIBUTION PLOT

General Notes

The programs in this section will give you a "feel" for many of the basic statistical concepts. They are easy to enter into any personal computer and can be used to experiment with variations in data to appreciate the consequences upon calculations. *However, it is important to note that these programs are not meant for formal use in data analysis.* They are not representative of the program strategies of more comprehensive statistical program packages available for larger computers or the newly available ones for personal computers. But like the skeletal data used in this book, they illustrate relations between data and selected statistical concepts.

The programs have been written in a version of the BASIC programming language that will operate on most home or personal computers, or if some adaptation is necessary, it will be minor. They have been written in the simplest possible form so that they will be very easy to enter and use.

So that they could be written simply, these programs typically follow the hand-calculator formulas for various statistics rather than the specific calculational steps in the text examples. However, even if you do not know BASIC, the calculations are relatively self-evident.

When entering programs, be sure to pay attention to distinctions between the keys for "l" and "1" or "0" and "O" because the computer depends upon this.

Mean, Variance, and Standard Deviation

This is a fairly common type of program that calculates the major distributional statistics for a set of scores. You might note that "N" rather than

"N − 1" is used here. (Use the latter when making estimates of population characteristics.) In trying this program, you might experiment with the effects of widely varying scores on the calculations of variance and standard deviation.

```
10 REM BASIC MEAN, VARIANCE, STANDARD DEVIATION
20 INPUT"NUMBER OF SCORES";N
30 FOR I = I TO N
40 INPUT"SCORE";X
50 A = A + X
60 B = B + X^2
70 NEXT I
80 M = A/N
90 V = (B − N * M^ 2) / N
100 PRINT
110 PRINT"MEAN = ";M
120 PRINT"VARIANCE = ";V
130 PRINT"STANDARD DEVIATION = ";SQR(V)
```

t Test

This is a program for an independent groups t test where the sizes of the two samples do not have to be equal.

```
10 REM t TEST
20 DIM X(100,2)
30 FOR I = 1 TO 2
40 A(I) = 0
50 B(I) = 0
60 PRINT"SAMPLE";I;":"
70 PRINT"SIZE OF SAMPLE";
80 INPUT N(I)
90 FOR J = 1 TO N(I)
100 PRINT"VALUE";J;
110 INPUT X(J,I)
120 A(I) = A(I) + X(J,I)
130 B(I) = B(I) + X(J,I)^2
140 NEXT J
150 M(I) = A(I)/N(I)
160 A(I) + (B(I) − A(I) ^2 / N(I))/(N(I)−1)
170 NEXT I
180 R = (M(I) − M(2))/SQR(1/N(1) +1/N(2))
190 S = N(1) + N(2) − 2
200 R = R/SQR(((N(1) − 1)* A(1) +(N(2) − 1) * A(2))/S)
210 PRINT
220 PRINT"MEAN 1 = ";M(1)
230 PRINT"MEAN 2 = ";M(2)
240 PRINT
```

```
250 PRINT"t VALUE = ";R
260 PRINT"d.f. = ";S
270 END
```

Chi Square

This handy little program will accumulate values of χ^2 as you enter pairs of observed and theoretical frequencies. (You'll have to calculate the theoretical frequencies on scratch paper beforehand.)

```
10 REM CHI SQUARE
20 INPUT"OBSERVED FREQUENCY";O
30 INPUT"EXPECTED FREQUENCY";E
40 A = ((O − E)^2)/E
50 B = B + A
60 PRINT"CHI SQUARE = ";B
70 GOTO 20
```

Correlation

This is a simple program for calculation of the Pearson product-moment correlation. Perhaps you can recall from your reading how correlation coefficients vary with small samples. You can demonstrate this effect by introducing a few variations as you try different sets of scores.

```
10 REM CORRELATION
20 INPUT"NUMBER OF PAIRS OF SCORES";N
30 FOR I = 1 TO N
40 INPUT "ENTER X,Y";X,Y
50 A = A + X
60 B = B + Y
70 C = C + X^2
80 D = D + Y^2
90 E = E = E + X * Y
100 NEXT I
110 R = (N * E − A * B) / N
120 S +(SQR ((N * C − A^2) * (N * D − B^2))) / N
130 PRINT"N = ";N
140 PRINT "CORRELATION = "; R/S
```

Regression

You can readily see that this regression program simply involves a few lines added to the foregoing correlation program. If you know some BASIC, you could also add lines for calculating standard error of the estimate.

```
 10 REM REGRESSION
 20 INPUT"NUMBER OF PAIRS OF SCORES";N
 30 FOR I = 1 to N
 40 INPUT"ENTER X,Y"; X,Y
 50 A = A + X
 60 B = B + Y
 70 C = C + X^2
 80 D = D + Y^2
 90 E = E + X * Y
100 NEXT I
110 R = (N * E − A * B) / N
120 S + (SQR ((N * C − A^2) * (N * D − B^2))) / N
130 PRINT"N = ";N
140 PRINT"CORRELATION =";R/S
150 PRINT
160 F = (N * E − B * A) / (N * C − A^2)
170 G = (B − F * A) / N
180 PRINT"PREDICTION:"
190 PRINT"Y = ";G;" + ";F;" TIMES X"
```

Distribution Plot

This simple program will generate a frequency distribution similar to the following:

```
23> *
24> **

26> ****
```

Note that scores missing from your array will result in blanks along the baseline (shown vertically) of the distribution. Because most computer screens are limited to about 24 lines, you will want to keep your range of scores within this figure. If your raw scores exceed this range, you could consider grouping them into categories (for example, entering "5" for scores in the 1 to 10 range, "15" for those in the 11 to 20 range). On the other hand, if you have a printer capability, you need not be affected by the screen size limitation. Simply substitute printout commands for the screen "PRINT" commands in the program. (Print commands vary for different versions of BASIC, for example it is "LPRINT" for the IBM.)

```
10 REM DIST
20 DIM X(100)
30 INPUT"NUMBER OF SCORES";N
40 FOR I = 1 TO N
50 INPUT X(I)
60 NEXT I
70 FOR I = 1 TO N
```

```
 80 FOR J = 1 TO N − 1
 90 X = X(J)
100 B = X(J + 1)
110 IF X < B THEN 140
120 X(J) = B
130 X (J + 1) = X
140 NEXT J
150 NEXT I
160 PRINT:PRINT
170 FOR I = 1 TO N
180 IF X(I) = X(I − 1) THEN 270
190 D = (X(I)) − (X(I − 1))
200 PRINT
210 FOR S = 1 TO D − 1
220 PRINT
230 NEXT S
240 PRINT X(I);">";
250 PRINT TAB(8);"*";
260 GOTO 280
270 PRINT"*";
280 NEXT I
```

INDEX

A

Analysis of covariance, 98–99
Analysis of variance
  calculation of F ratio, 93–94
  comparing individual means, 95–97
  F ratio in multiple-factor model, 106–113
  general considerations of F ratio, 95
  levels of a factor, 98
  logic underlying F ratio, 93–95
  multiple-factor model, 101–115
  multivariate analysis of variance, 99–100, 115
  random compared with fixed factors, 97
  relationship with regression, 156
  repeated measures, 98
  sample study involving, 90–93, 101–106
  single-factor model, 90–100
Arithmetic mean. *See* Mean

B

Beta coefficients, 160
Between-groups variance, 93
Binary variables, 142–143, 155–156, 165–166
Binomial expansion, 56

C

Canonical correlation, 196
Categorical variable, 24
Central tendency, 38–40
Centroids, 196
Chi-square
  calculation of, 120
  computer program for, 212
  differences among samples, 119–120
  interpretation of, 121
  log-linear models of, 121
  one-sample case, 117–119
Chi-square table, 208–209
Classificatory scale, 23–25

Cluster analysis
  description of, 171
  practical points concerning, 175
  sample study involving, 171–174
Coefficient of correlation, 131–132
Coefficient of multiple correlation, 141
Coefficient of partial correlation, 144
Commonality, 188–189
Computer programs, 210–214
Confirmatory factor analysis, 192
Constants, 152
Continuous variables, 29
Control variable, 99
Correlation
  binary or dummy variables in, 142–143
  coefficient of correlation, 131–132
  computer program for, 212
  logic underlying product-moment correlation, 134–142
  multiple correlation, 140–142
  nature of, 131–132
  negative correlation, 131
  nonparametric approaches to, 144–145
  partial correlation, 143–144
  positive correlation, 131
  sample study involving, 132–134
  significance of, 132
Correlation matrix, 185–186
Covariate, 99
Criterion variable, 99

D

Data (plural), 11
Degrees of freedom, 87–88, 121
Dependent variable, 15
Descriptive method, 15
Descriptive statistics, 13, 43–45
Discriminant analysis
  discriminant scores and probabilities, 202–203
  example from demographics and media uses data, 203–204
  interpretation of coefficients, 200–201

215